Kombucha – Miracle Fungus

THE ESSENTIAL HANDBOOK

Kombucha – Miracle Fungus

THE ESSENTIAL HANDBOOK

Revised Edition

by Harald W. Tietze

Gateway Books, Bath

Published by
GATEWAY BOOKS,
The Hollies, Wellow,
Bath, BA2 8QJ, U.K.

Copyright © 1994 Harald W.Tietze

First edition May 1994 as Kombucha – The Miracle Fungus
Revised reprints August 1994, March 1995
Second edition (revised) June 1996 as
Kombucha – Miracle Fungus

Reprinted October 1996

Distributed in the U.S.A. by
NATIONAL BOOK NETWORK
4720 Boston Way, Lanham, MD 20706

Cover design by Synergie, Bristol
Photograph courtesy Kevin Redpath

Text set in Times $10^1/_2$ on 13pt
by Character Graphics, Taunton
Printed and bound by
Redwood Books of Trowbridge

British Library Cataloguing-in Publication Data:
A catalogue record for this book is
available from the British Library

ISBN 1-85860-042-1 (revised)

CONTENTS

FOREWORD

Every morning we start the day with a refreshing tea-drink. Nothing unusual about that you might think, except that our tea is no ordinary brewed beverage. It is Kombucha, the revitalising and delicious drink which has deservedly earned itself the reputation of 'miracle elixir'.

Kombucha is an ancient, fungus-like living organism which, when brewed with sweetened tea, produces a wonderful, health-enhancing drink. We were introduced to this immune-boosting treatment through a cousin in California, the sole survivor in a family who had all contracted cancer through a polluted public water supply. Kombucha encourages sharing, for it reproduces an offspring every week, and one finds someone to give it to. We brought ours back to Britain in June 1994, and have since given away hundreds to our publishing colleagues, authors, family and friends and through The Kombucha Network. Most have adopted it successfully, reporting definite and individual improvements to their wellbeing and health.

Its use quickly becomes a way of life, enabling us to do something practical at home for our good health and for our family, costing virtually nothing. We can then have the added pleasure of passing Kombucha on to our friends and neighbours, also for their good health, happiness and wellbeing. It is easy to understand why its production and consumption in rural Russia is a way of life, and is passed on from one generation to another. Making Kombucha is simple and regarded by many as a sacred ritual, by others as a focused time when they put care and consideration into this wonderful living drink.

We have now heard that the Aids community in California is taking Kombucha as an antidote to the side-effects of AZT, and its fame has rapidly grown as a marvellous health drink. The British media have noted its rising popularity in the UK. At a time when the amount of pollution in our environments has grown alarmingly, our immune systems are going to need all the help they can get. What could be better than a natural source of immune enhancement with a 2,500 year pedigree of healing — and no bad press at all!

This book has an interesting history. It was first published in Australia in 1994 and circulated initially for the German speaking community there. Since then it has taken the Australian, British and American health markets by storm. The first edition in English has sold more than 100,000 copies and is now in its sixth printing.

In this new edition there is more information on how to produce the Kombucha drink, advice on varying the taste and adding herbs for improving the therapeutic effect. Anecdotal evidence of how Kombucha therapy has helped people continues to mount and surprise us daily. It would be misleading to call it a panacea for all ills, but many individuals have told us how they have received specific help from drinking it, often in ways they didn't anticipate.

This very informative and practical book covers all the interesting and relevant points you are likely to want to know. Knowledge is important and empowering, giving confidence and clarity to the brewing and sharing of this potent and remarkable beverage.

Good Health and enjoy!
Alick and Mari Bartholomew, Wellow, March 1996.

INTRODUCTION

Kombucha is an ancient food and healing source originating in Asia. Thanks to a growing interest in Eastern healing methods, Kombucha travelled via Russia into Western Europe. In those countries, Kombucha was regarded as a pleasant health-giving beverage which contributed to a balanced diet. On its way to the Western World this potion has often appeared to have performed miracles. It acquired names such as magical fungus, miracle fungus and elixir of long life because of its success in the treatment of modern illnesses. While Kombucha was already a popular healing remedy in the 1920s and 1930s in Russia, Czechoslovakia, Germany and Austria, the fermented drink almost disappeared during and after World War II. This was because sugar and tea were in short supply.

In the 1960s Kombucha was brought into Germany and Austria from Eastern Europe. In the 1970s, when the standard of living improved, it became known again as a miracle beverage; the trials and medical research into its effects were conducted mostly in Germany and were published in the German language. Articles about it appeared in many newspapers.

In modern societies, with their abundance of sterilised, homogenised or preserved food, a microbiological digestive aid such as Kombucha can work wonders. Many doctors and naturopaths recommend Kombucha-drinking for all types of illness.

Kombucha, like many other products of natural medicine, should not be seen as a cure-all; there is no such thing. Health is the balance of body, emotions and mind. When an illness is cured it is because the cause has been found and subsequently the symptoms disappear. A health beverage like Kombucha can only be of optimum benefit if there is also balance in our diet, lifestyle and attitudes.

Among its main claims, Kombucha strengthens the immune system, so helping in a cure for many illnesses. It also detoxifies, cleansing the blood and strengthening the kidneys. Kombucha also plays a vital role in regenerating bowel flora, and is a good aid to the whole digestive system. It is exciting news that Kombucha is gaining in popularity and is being recognised as a remarkable life-enhancer. This wonderful living organism produces a refreshing, revitalising and delicious drink, easily made at home by all the family. Kombucha is a gift of Nature that can be passed on from friend to friend and from one generation to another. Look after Kombucha and Kombucha will look after you!

DISCLAIMER
This publication seeks to demonstrate the beneficial use of Kombucha for various disorders in the human body, but it is stressed that the contents of this book are in no way a substitute for personal supervision. People with health problems should consult their health practitioner.

Part I:
WHERE DOES KOMBUCHA COME FROM?

ORIGIN OF THE KOMBUCHA FUNGUS

No one can say, for sure, how and where the Kombucha fungus originated, but we do know that it has been used for at least two thousand years. It does not really matter whether the Chinese, Koreans, Japanese or Russians were the first to ferment the fungus. To be exact, it isn't really a fungus, as such, but rather a community of yeast and bacteria. One authority describes it as a lichen. The origin of the name could be Japanese, with 'kombu' standing for the brown tea algae and 'cha' for tea. The best description, in my opinion, is that of the herbalist, Pastor Weidinger, whom I have used as a source throughout this book. He says:

'Kombucha' tea is an ancient East Asiatic beverage which came out of the ocean. For three years I was a missionary on the island of Taiwan. This south-eastern coastal region with its subtropical climate and extended growing period made ideal conditions for growing the tea, which was farmed in large areas. This province is regarded as the origin for 'tea', ie. in Latin 'Thea'. The original name given by the Chinese writer Kuo-Po to the beverage extracted from these leaves was 'Tu' or 'Tschuan'. Today it is called 'Ch'a'. In the province of Fuiken, however it is still called 'T'e'. This has given me a closer understanding of the word tea. K'un-Pu-ch'a, a tea-like wine......

My missionary activities also required my travels to the islands of Quemoi and Matsu which are situated close to the mainland near the Province of Fukien. I was very impressed by a beverage which was served to me by the local people which had a sweet-sour taste and was very refreshing in the hot climate of the area. Was it wine which tasted like a delicious tea or an unusual tea that tasted like a rare wine? Notably, after drinking this beverage, I not only felt stronger after the long and tiresome walk, but strangely enough, felt healthier. In particular it helped my metabolism a great deal in this climate and made me feel very relaxed. When I asked what it was, "K'un-Pu-ch'a" came the reply. I was startled. "Tea which came from life in the ocean"? Already in the Tsin-Dynasty, about 221 BC, it was known and hon-

oured as a beverage with magical powers enabling people to live for-ever. The tea was given different names. One of the most famous was the 'Godly Tsche'. This particular tea was used as a remedy for chronic gastritis. People also tell of the Korean medicine man named Kombu who in the year 414 prescribed the tea to heal the Japanese Emperor's disorders. The 'Godly Tsche' came from China, via Korea to Japan where it was given the name 'Tsche of Kombu'.

KOMBUCHA IN THE WESTERN WORLD

The beverage eventually reached Europe via Mongolia and Russia. When I mentioned Kombucha in an article published in Australia dealing with traditional medicine, I received countless letters from people who originated in the old eastern parts of Germany and who now live in Australia. They recalled their grandmothers giving them a refreshing beverage from a large pot, on top of which floated a type of fungus. Others remembered the recipe being brought by Polish workers from the East. In World War I the fungus-tea disappeared because of the sugar shortage.

More reports about the existence of this magical fungus began to emerge in the 1920s when researchers started to show an interest in its properties. Their studies were suspended with the beginning of World War II. Kombucha only really became popular again in the 1960s and 1970s, probably as an antidote to unhealthy eating habits. The Western media took up the topic. Reports came out of Kargasok in Russia of people who reached a great age. This phenomenon appeared to be linked with the drinking of Kombucha, which is why it is also known as 'Kargasok Tea'. According to a report, an 80 year old woman gave birth to a healthy baby, fathered by a 130 year old man!

The most famous research results come from the University of Omsk in Russia and, in the West, from the research of Dr Rudolf Sklenar. Dr Sklenar's research is often mentioned in the German press. He came from Eastern Germany where Kombucha has been used since the turn of the century among ordinary people. He studied medicine in Prague and had his first contact with the Kombucha fungus in a monastery. He worked with the culture during the second World War and based his scientific work upon Kombucha. In the 1960s he published his research in the scientific and general press, which resulted in increased awareness of Kombucha's healing and strengthening properties. Dr Sklenar used Kombucha successfully for diabetes, high blood pressure, all types of digestive problems, stomach and bowel illnesses, rheumatism and gout.

Dr Sklenar's main area of work became the biological treatment of cancer, and he integrated Kombucha with this programme. His healing methods were so successful that they were adopted by many doctors. A company which carries his name today manufactures Kombucha beverage and Kombucha drops. When it was reported that Dr Veronika Carstens (wife of a former German President) was using Kombucha with all her cancer patients, the fungus beverage became a popular healing remedy. Dr Sklenar and Dr Carstens recommended the use of Kombucha to their cancer patients as a complement to other therapies.

When it became known that the U.S. President, Ronald Reagan, was suffering from cancer, reports regarding his treatment were lost in the larger issues of politics. According to researcher and author, Gunther Frank, Ronald Reagan heard of Kombucha through the autobiography of the Nobel Prize author Aleksandr Solzhenitsyn. He was diagnosed as having cancer in 1952 and fully recovered at the hospital in Tashkent ('The Cancer Ward') in 1953. Ronald Reagan received a culture from Japan and reportedly drank a litre of Kombucha daily. His cancer was prevented from spreading further and Reagan was able to finish his term in office. Dr Robert E.Willner (USA) cites Kombucha as being helpful in his highly recommended book *The Cancer Solution*.

Immigrants from Asia and Eastern Europe introduced the Kombucha fungus to the Western world. Today, a number of health practitioners who have treated their patients for all types of illness with the Kombucha beverage have found that, not only did their patients feel much better after drinking it but that, when used in conjunction with other medicine, the healing results were greater.

A Kombucha pioneer in Australia, Mr Jose Perko from Kombucha House in Queensland, believes that Kombucha made with green tea has better therapeutic value than that made with the black tea which is generally used in other countries. By experimenting with different fruit and herbal mixtures, (one of which is a pawpaw leaf concentrate), Mr Perko has managed to keep Kombucha mixtures for long periods without the use of artificial preservatives or causing unnecessary souring of the Kombucha. In further chapters I will discuss in detail some of his methods. I would like to thank him for the information he has supplied for inclusion in this book.

THE MANY NAMES OF KOMBUCHA

There are many names for the Kombucha fungus and for the brewed drink made from it. An interesting fact is that many of these names incorporate the word fungus, even though it is not really one. I call it Kombucha fungus and regard it as just a name rather than a botanical description. Other commonly-used names range from algae fungus to magical fungus.

In various languages names such as 'Russian flower', 'Russian jelly', 'Russian fungus', 'Japanese fungus', 'Japanese sponge', 'Russian mother' and 'Indian wine fungus' point to its heritage. 'Wondrous fungus', 'Magical fungus', 'Heroic fungus', 'Fungus giving long life' and 'Gout jelly' are names also given to the fungus in various languages.

Kombucha has the effect of giving the recipient a general feeling of wellbeing and is usually considered to be a remedy for many illnesses. With the beverage itself, most of the names are linked to the actual taste, such as 'Tea beer', 'Tea wine' and 'Tea cider'. In France it has been given the name of 'Elixir de la longue vie' - elixir of long life - which once again links the name to the effects it has on the health of Kombucha drinkers.

Whether Kombucha is spelt with K or C doesn't really matter. Due to the intensive research conducted by famous doctors such as Dr Sklenar, Dr Wiesner and Dr Carstens, it appears that the name Kombucha dominates. The German bioculture expert, Dr Alex Meixner, calls it Combucha in his book *Pilze selber zuchten (How to grow mushrooms);* at the same time he issues a warning that all Kombucha products are not of the same quality.

A list of all the names I have found so far follows:
Algae fungus; Algentee; Brinum Ssene; Cajnogo griba; Cajnyj grib; Cajnyj kvas; Cembuya orientalis; Chamboucho; Champagne of life; Champignon Japonaise; Champignon Chinois; Champignon miracle; Champignon de la Charite'; Champignon de longue vie; China-Pilz; Chinesischer Teepilz; Ciuperca de ceai; Comboucha; Combucha; Combuchaschwamm; Combuchatee; Conbucha; Elixir de longue vie; Devine Che;

Fungojapon; Fungus japonicus; Fungus tea; Funko Cinese; Ganoderma japonicum Gichtqualle; Gift of life; Godly Tsche; Gout fungus; Grib; Haipao; Heldenpilz; Hero fungus; Hongo; Indian wine fungus; Indian mushroom; Indischer Weinpilz; Indischer Teepilz; Japa'n gomba; Japanese fungus; Japanese sponge; Japanischer Teepilz; Japanischer Combucha; Japanisches Mutterchen; Japanpilz; Japonskagliva; Jponskij grib;

K'un-Pu-ch'a; Kambuha; Kargasok Pilz; Kargasok Schwamm; Kargasoktee; Kocha Kinoko; Komboecha; Komboecha-drank; Kombucha Elixir; Kombucha tea fungus; Kombucha; Kombucha-getrdnk; Kombuchamost; Kombuchawein; Kombucha-thee; Kombucha-schwamm; Konko; Kwas; Kwassan; Lingzhi; Magic mushroom; Manchurian Elixir;

Manchurian fungus; Manchurian tea; Mandschurisch-japanischer Pilz; Mandschurischer Schwamm; Medusentee; Medusomyces Gisevii Lindau; Miracle fungus; Mo-Gu; Olinka;

Red tea fungus; Reishi;   Russian tea-vinegar; Russian flower; Russian fungus; Russische Qualle; Russische Blume; Russische Mutter; Russischer Pilz; Sakwaska; Symbiont Schizosaccharomyces Pombe-Bacterium xylinum; Tea fungus Kombucha; Tea fungus; Tea cider; Tea beer; Tea kwas; Tea mould; Tea plant; Tea wine; Teemost; Teekwa; Teewein; Teepilz; Teeschwamm; Teyi saki; Thee-Schimmel; Theebier; Theezwam; Tschambucco; Tsche of Kombu; Wolgameduse; Wolgapilz; Wolgaqualle; Wunderpilz; Yaponge; Zauberpilz; Zaubersaft; Zaubertrank.

WHAT IS KOMBUCHA?

Kombucha is composed of a number of bacteria and special yeast cultures (different to those affecting candida suffers — see p.78) in a symbiotic relationship. This living organism ferments sweetened tea to become the Kombucha beverage. The refreshing taste gives one a feeling of wellbeing. The amount of living yeast it contains gives the Kombucha beverage an active life which continues after it is decanted into bottles. Kombucha has become increasingly highly regarded as an aid to preventing many illnesses.

KARGASOK TEA

Reports dealing with healing successes are often generalised and passed on without the desired research. When I first received the Kombucha fungus from a friend, I was given an information leaflet about Kargasok tea. Unfortunately, I have not been able to trace the author, but I will pass the report on word for word:

"Approximately 60 years ago, a Japanese woman visited the region of Kargasok (in Russia) and was stunned to find so many healthy people who were over the age of 100. She actually met a man at the age of 130 who had married an old woman of over eighty who was still able to conceive children! The Japanese woman was fascinated by this and attempted to obtain the secret of the eighty-year-old woman who hardly had a wrinkle. She discovered that in every household, young and old alike consumed about 1pint (US) or one-third litre of Kombucha tea each day. To make this tea, the Japanese woman received a special yeast fungus and instructions on how to use it. She took the fungus with her to Japan where she started to duplicate it. Her friends were invited to drink the tea as well and she also passed the fungus on to them with instructions on its use. They passed it on to their friends. After having consumed the tea for a period of time, the people began reporting successful effects :

- *A man with a blood pressure of 210/120 was able to reduce it to 140/80.*
- *A young girl diagnosed with shingles was cured.*

"In Kargasok, cancer and high blood pressure are unknown. In Japan, soon after this tea became the subject of TV and radio programmes, well over one million Japanese were consuming the tea. Eventually the tea found its way to Taiwan, then to Hong Kong and now travels around the world where it is passed on from one friend to another as a token of appreciation and love. This tea appears to be a miraculous remedy for many types of suffering. Researchers have discovered that the fungus has three basic elements, without which the body cannot function. Encouragingly, Dr Pan Pen from Japan reported as follows on the effects of the culture: 'This tea —

- *clearly lengthens the lifespan*
- *is a health remedy against chickenpox and shingles*
- *reduces the formation of wrinkles*
- *discourages the formation of cancer*
- *prevents adverse menopausal symptoms*
- *restores visual acuity*
- *strengthens leg muscles*

- *heals arthritis*
- *enhances sexual drive*
- *heals sweaty feet, constipation, joint and back pains*
- *heals abscesses*
- *heals blocked arteries and diabetes*
- *strengthens kidneys*
- *heals cataracts and heart disease*
- *restores the appetite and heals sleeping disorders*
- *reduces the chance of gall stones and liver problems*
- *reduces obesity and stops diarrhoea*
- *heals haemorrhoids*
- *helps restore colour to grey hair and improves baldness'"*

These results may appear in certain individual cases, but one should not generalise from this that Kombucha will always help those conditions. There was no record of the type of teas that were used, nor whether the tea brew or the fungus itself were being used. It would be unusual, in my view, for example, for wrinkles to be prevented just by drinking the beverage. However, rubbing the pulp on to the face could certainly lead to improvements. As to the healing of sweaty feet, the restoration of color to grey hair and the prevention of baldness, I am not convinced. It should be determined whether grey hair or baldness are the result of an illness or just genetic. My hair has started to turn white, as is usual in my family and, although I have used Kombucha for a number of years, it has not restored my original blond hair. I cannot claim, either, that my chronic travel and motion sickness have been prevented by the use of Kombucha. However the results in relation to wellbeing and a good digestion can certainly be connected to the use of Kombucha.

Can Kombucha be considered a miracle healing medium? Is it a remedy against cancer? Research will undoubtedly determine this in the future.

Part II:
HOW TO MAKE YOUR OWN KOMBUCHA
Everything You Need to Know

THE BASIC RECIPE:

INGREDIENTS:

USA

10 cups (5 US pints) of water (filtered or uncarbonated bottled for preference - see p.26)

³/₄ cup white, granulated sugar*

2 teaspoons (or teabags) tea — black, green, or a mixture

1 healthy Kombucha fungus**

2 tablespoons cider vinegar or white vinegar (This is used only with the first brew, if no starter (mother) Kombucha tea is available. In subsequent brews, use 1 cup (¹/₂ pint approx.) of the mother brew as a starter and omit vinegar.)

Metric/Imperial

2 litres (4 pints) of water (filtered or uncarbonated bottled for preference - see p.26)

160 grams (5¹/₂ oz) of white, granulated sugar*

2 teaspoons (or teabags) ordinary black or green tea

1 healthy Kombucha fungus**

2 tablespoons cider vinegar or white vinegar (This is used only with the first brew, if no starter (mother) Kombucha tea is available. In subsequent brews, use 200ml (¹/₃ pint) of the Kombucha brew as a starter and omit vinegar.)

We recommend you start brewing with black or green tea; when you are more confident you can experiment with herbs. (See Part IV,*"Making the Beverage more Interesting"*, pp.56-62)

---

*Reduction in the amount of sugar recommended or substitution of brown sugar could prejudice the health of the fungus.

**To find out how to obtain a healthy Kombucha fungus see "Appendix Two" on p.105.

+ sugar

Fig. 1. Make sweetened tea in a teapot or bowl

Container: china, glass or porcelain (approx. 6 pints vol.), large surface area for fast fermentation!

Fig. 2. Strain sweetened tea into suitable container. Top up with remaining water

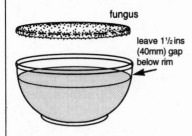

fungus

leave 1½ ins (40mm) gap below rim

Fig. 3. Add vinegar or starter brew. Place fungus on top of tea, smooth side up.

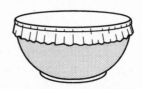

Fig. 4. Cover with muslin or other suitable fabric and secure with elastic. Place in a warm dark place away from dampness and plants to ferment

Separate new layer from original fungus.

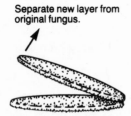

Fig. 5. After fermentation is complete separated 'offspring' may be used for new brew, stored in fridge or passed on to a friend

Fig. 6. Kombucha brew now ready to drink. Strain into glass bottles or jugs and store in fridge – enjoy!

METHOD

(From now onwards the amounts given are for the U.S.A. with metric/ Imperial measurements in brackets): To make a batch of Kombucha, you will need a 15 cup (3 litres/ 6 pints) sized bowl. Place the tea in a large teapot or bowl, pour on boiling water, add sugar, stir until dissolved, and leave to brew for 10 minutes. Remove tea-bags or strain brewed tea into the bowl. Add the remainder of the water to bring the total amount to 5 US pints (2 litres /4 pints) and allow to cool to room temperature. Add the starter brew (or vinegar as a starter the first time). Now place the fungus in the liquid. It will eventually float, but don't worry if it sinks at first! The fungus has a smooth side (possibly lighter in colour) and a rougher side; place the smooth side upwards. A gap of at least one and a half inches should be left between the fungus and the top of the bowl.

Cover the container with muslin (or some other cloth of natural fibre which will allow air through) and anchor this below the lip of the bowl with elastic. (This stops insects getting into the fermentation container.)

The container should then be stood in a warm place or use a heated brewing mat. The ideal environmental fermentation temperature is 70° - 84°F (23° - 28°C); use a simple thermometer to check it. The fungus does not require any light, but needs warmth and air. Tobacco smoke is harmful. After 5 - 10 days of fermentation (this is faster in summer or in higher temperatures) your Kombucha brew will be ready. But you do need to start tasting it to see how it is doing, after 4-5 days. It should taste neither 'flat' nor too sweet — if it does, then the sugar has not been converted into healthy organic acids. So cover and leave for another day or more. When the brew tastes zingy and not sweet but with a pleasant cider-like tang to it, then it is ready. Now remove the fungus with clean hands onto a plate. You will notice that a new layer has formed on top. Strain the fermented beverage and pour into bottles, leaving a little air space at the top. Alternatively, the beverage can be poured into one or more lidded jugs. The bottles or jugs should be placed in the fridge, or other cool place otherwise the fermentation process will continue, and the beverage will start to turn sour. Once the fungus has been removed from the tea-brew, and the off-spring separated, the original fungus can immediately be used to start a new batch. The refrigerated brew is now ready to drink.

THE FUNGUS REPRODUCES ITSELF

With each brew, a new fungus will have grown on top of the original (by binary fission). The two layers can be parted easily, (if in difficulty, use a clean pair of scissors to gently ease them apart). This new fungus can now be used to start another batch or passed on to a friend. If no batches are started on the same day, the 'baby' fungus can be preserved by placing it in an air-tight container covering it with some Kombucha beverage. Leave an air space between the liquid and the lid and keep it in the fridge until required. The recommended amount to drink daily is one average-sized wineglass, approx. $^1/_2$ US cup (150mls/$^1/_4$ pt three times per day), one before breakfast and one with or after each other meal. Larger amounts can be consumed quite safely. (Important: You may find it more beneficial to start drinking Kombucha in smaller amounts, gradually building up to the full recommended amount — see p.36 — *Dosage: How much can one drink?*)

THE RIGHT BREWING TEMPERATURE

The ideal fermentation temperature is 70°-84°F (23°-28°C), which is higher than the average temperature in most parts of the world. It does not matter if the temperature varies from a low of 14°C (57°F) at night to over 84°F (28°C) during the day. A room temperature of over 97°F (36°C) does not mean that the brew itself would have the same temperature. A very high temperature will cause high evaporation; some people in the tropics report a 25% loss in the brew volume during fermentation. This concentrating effect has to be considered when brewing medicinal herbal teas as a treatment for illness. In the first days of fermentation the right brewing temperature is important to avoid contamination. In the cooler months a brewing time of 10 days or more may be needed to ferment the brew to a health-giving drink. One should not go by the brewing time only. In different areas and at different times of the year adjustments have to be made.

SOME OTHER HEATING METHODS:

Airing Cupboard: This is ideal, especially when fitted with slatted wooden shelves. Do leave the door slightly ajar to allow air to circulate.

Heated Trays: Sold in two different sizes in home brewing stores and by mail order. *See Appendix Two pp. 105 and 107.*

A typical two litre bowl with muslin cover secured by elastic

Heating belt: Designed to wrap around the container – again sold in brewing departments.

Submersible heater: The kind used for fish tanks – suitable for big brewing containers.

THE BREWING POT

The most suitable containers for brewing are made of glass, porcelain or high glazed earthenware. Metal containers should not be used because acids in the brew will react with the metal. A Kombucha brew made in a stainless steel container does not taste so good as one made in a recommended type. If plastic is used, it should be of the highest quality food grade and acid-resistant. Polyvinyls, polypropylenes and cheap plastics can cause chemical reactions in the brew. To grow quickly, the culture needs a large surface area, so the container should have a wide opening and not be too tall. It should not be filled to the top. A wider, shallower pot allows quicker, better fermentation.

For the Continuous Fermentation Method a beer brewing container used for home brewing may be used, though a high glazed authentic ceramic jar is best and is, of course, very attractive. *(See Appendix Two Page 105).*

WATER – CHEMICAL-FREE PLEASE

One of the main causes of poor health is poor water. Good, clean chemical-free water is a basic necessity in any programme to improve health but very few modern communities have it. Contrary to what water authorities say, there are still undesirable elements in our tap water, and even the maximum safe limits recommended by national and supra-national authorities are not being met by all water companies. These undesirable elements are chemical in nature. Some – like chlorine, aluminum sulphate and fluorides – are added on purpose to our supplies. Others, such as pesticides, herbicides, fertilisers and metals, can leach into our underground water as a result of industrial pollution.

Chlorine is added by water companies to kill bacteria, particularly in ancient pipework serving cities. As Kombucha contains millions of friendly bacteria, it is important to brew with either good quality uncarbonated bottled water or efficiently-filtered water. Though some jug systems are fine, an under-the-sink plumbed-in type of filter with

A traditional pot for continuous fermentation. Note the tap is above the yeast setting level.

its own tap is the most convenient for all your drinking and cooking needs. *(see Appendix Two for suppliers)*

CONTINUOUS FERMENTATION PROCESS

After I published my first booklet about Kombucha, I received quite a number of comments and enquiries from people of East European origin. These people recalled their mothers or grandmothers having a large container, stored in a warm place. People drank the beverage when they wanted to, every now and again topping up the container with sweetened tea. This was continuous fermentation.

Continuous fermentation has many advantages over Western batch brewing. Besides being easier and faster, it has the advantage that a larger amount of fluid does not react so quickly to temperature variations. A large container can also be used with an electric heating tray *(See Appendix Two for suppliers)*. In my experience, continuous fermentation is the ideal way to brew Kombucha for personal and family use.

To start, you need a big container, approximately 5 US galls (20 lit/40 pts) suitable for Kombucha brewing. A home brew fermenter used for beer brewing will do with a tap located several centimetres above the base, to avoid the yeast sediment in the bottom. A glazed porcelain container is even better, and is also decorative. Start with at least 10% starter brew from the previous batch as you do with batch brewing, and multiply the tea, sugar and water in proportion to the amounts you use in your batch container. If you have more starter brew available, use as much as you have. When you start with 10 litres, for example, you multiply the given recipe on page 21 by 5. Pour the cooled sweet tea into the fermenter. Add the starter brew and a piece of culture. A smaller piece would be adequate. It does not matter if the culture sinks to the bottom. The new culture will develop on the surface of the brew. After 6 - 10 days, depending on temperature and the amount of starter brew used, you should begin tasting, pouring from the tap on the container. Start drinking when the brew tastes zingy and cider-like.

Topping up is necessary when between 10%-30% of the brew has been drunk. This amount can be drawn off in one go and stored in the refrigerator. Top up with sweet tea and sugar in the right proportions. *Don't add hot tea to the fermenter,* let it cool first to at least room or body temperature. For topping up, different teas can be used every time, according to taste or as a health stimulant. Children love a mix with wild blackcurrant, rosehip and other fruit teas. With the range or different top-ups available you will discover the wonderland of Kombucha brewing.

Maintaining the continuous fermentation process is simple and your fermenter will teach you as you go along. Cleaning out sediments is necessary one or twice a year. Drain the pot to clear out the excess sediment when the tap, which should be several centimetres above the base, is blocked. If you use additives like crushed ginger directly into the brew, cleaning of the sediments might be necessary every 2 - 3 months. Don't take the culture out of the pot if there is no reason to do so. Many people wash their culture every week which is quite unnecessary. There is more danger of contamination in this way and applying chlorinated water to the culture may be harmful. The only time a culture should be 'washed' is if a mould has developed on the top. In this case it is better to throw the culture away and use a spare one, cleaning all equipment thoroughly and starting again. Washing the Kombucha culture with lemon juice or vinegar should be done only if there is no other culture available.

Separating the new cultures as they grow every week is another unnecessary task. Leave the culture alone and let it work without interference. The only time you need to take out the culture is when it takes up too much of the fermentation space. If your culture grows too big, take it out, removing a piece to put back in the fermenter. It doesn't matter if the culture is upside down after topping up. It will always grow a new one on the surface.

In Russia and Poland, Kombucha brewing is kept a secret from the children and they are not allowed to look inside the fermenter. The culture looks ugly and unappetising, and they might be put off drinking it. Diabetes and Candida sufferers should bottle some of the brew when it is well fermented and drink only this. They definitely should not drink from the fermenter the morning after topping up. If you feel your culture is not performing properly, top it up with a bottle of good quality commercial beverage or beg some from a friend who is brewing well.

SOME FREQUENTLY-ASKED QUESTIONS ABOUT BREWING KOMBUCHA

How do I know if the fungus is working? Don't be surprised if it sinks to the bottom; it may even stay there for a few days. The fungus may have previously been stored in the donor's refrigerator and is in semi-hibernation. The carbon dioxide generated in the warm conditions and the reactivation of the fungus will bring it back to the surface in time, don't worry. Even if the fungus remained on the bottom the whole of the first fermentation, it would still produce a new 'baby' on the surface of the liquid, the size of your bowl and will have a different colour depending on the tea used (eg brown from black tea, white from green tea).

Does it matter what size and shape the fungus is? No, as long as the fungus comes from a healthy parent that has been producing well. A piece 2-4 ins (5-10cms) square will work fine. Sometimes a new fungus, removed from a 'parent', has a hole caused by a gas bubble or is torn in separation. This imperfectly shaped culture as well.

What happens if I go on vacation for a while? The Kombucha fungus will remain in a dormant state quite happily for 3 - 6 months, if covered in Kombucha brew and left in a lidded container in the refrigerator. It likes an air space between the surface of the fungus and the lid.

How many times can I use my original fungus for batch brewing? About 5 - 6 times; after this it doesn't work quite so well. Always keep a spare healthy offspring in the refrigerator to start a new batch.

I am told the fungus needs to be smooth side up; does it matter which way round it is? I have tried it both ways and the fungus still works and produces a new layer. However, the fungus collects a brown film and loose bits on the underside. It is preferable to keep the fungus smooth side up in future batch brews.

Does the fungus need air? Yes, if you brew Kombucha in a warm cupboard, leave the door slightly ajar to allow air to circulate. Don't cover the bowl with a lid, but use only a cloth that will allow air through, securing with elastic. Without oxygen in the fermentation process the fungus will not work, and the unhealthy conditions can produce a mould.

Do I have to remove rings from my fingers when handling the fungus? This is unnecessary. Also, you can cut your fungus with clean scissors and boil the water in a kettle with a metal electric element. It is in the fermentation process that you must not use steel or an aluminium pot. (Aluminium is not recommended for any cooking).

Can pregnant and lactating women drink Kombucha? It is advisable to first contact their health practitioner before drinking Kombucha. Avoid caffeine during this time, so instead use herbal teas, eg. raspberry leaf (in moderation), dandelion, elderflower or lemon grass. (Avoid pawpaw as this can cause an abortion.)

What happens if the Kombucha brew tastes sour? You have left it to brew too long. It should still be drinkable and will do you no harm, and you can always add water or juice to make it palatable. Next time check earlier to see how the brew is fermenting. Ideally it should be neither sweet nor sour when it is ready for consuming.

Does Kombucha have to be in the dark during fermentation? Light is not required for the process of fermentation, and can be a hindrance. It has been shown that the yeasts and micro-organisms in the Kombucha culture are sensitive to sunlight as well as to ultra-violet light. So do ensure that your brewing container is kept in a shaded or dark place.

N.B. Kombucha brewing is a very simple brewing process from the East. So often when things come from the East to the West, people make a science out of simplicity — please beware of flyers and publications with inadequate or misleading information.

PROBLEMS ASSOCIATED WITH KOMBUCHA FERMENTATION

With any living fungus, accidents can occur which may prove harmful or even lethal to the organism. If the temperatures are not regulated or if the fungus has been stored for a long period of time, it can become more slimy than usual. If this happens, the fungus should be cleaned, using lemon juice or wine vinegar and can then be used again. Don't wash the fungus if there is no reason for doing so. It is always a bit slippery. In one case I came across a fungus which was not working properly even though its carer was doing everything correctly. I sought out the person who'd supplied the fungus and found that his 25 litre container was filled to the top, had a narrow neck, and the fabric cover was touching the fungus. It simply did not have enough air to perform properly. Even after doubling the fermentation time the brew was too sweet and the drink unhealthy. I traced the culture back further to the previous supplier where it had not done well either. Brewing had been done here for the previous 5 months with only 20grams of sugar per litre instead of 80grams. All the cultures have now been brought back to normal performance. With two of the cultures I added commercial beverage, and with the other I used double strength black tea with double sugar and fermented until it became very sour. From this brew normal fermentation was continued with good results.

It is a little harder to correct if the fungus has started to grow mouldy. Too low temperatures or an unclean environment can be the cause of this. In this case, dispose of the beverage and wash the fungus as described above. The usual 10% should be taken from another brew to start a new batch. If there are no other batches available, use vinegar as a starter as in your first brew (preferably Kombucha vinegar). The beverage should have a fruity, vinegary smell after a few days. This is a good indication that the fungus is working. It shouldn't have a mouldy smell. But to be safe, the best course of action is to dispose of a mouldy fungus and start with a new healthy one, correcting the problem that was the cause.

In areas where the vinegar fly *(Drosophila fenestarum)* is active, it is very important to cover the beverage with a muslin cloth fastened by an elastic band to prevent the fly entering the container and laying eggs. *(Meixner)*.

NB: A dead or ailing culture cannot perform properly and will not produce a healthy drink.

NICOTINE: A DEADLY POISON FOR THE FUNGUS

Tobacco smoke will kill the fungus. People have often contacted me and said that their fungus was not working any more. In 50% of these cases I found that someone was smoking in the room where the fungus was fermenting. The fungus cannot tolerate tobacco smoke. With occasional chimney smoke the fungus may survive. The beverage will, however, take on a smoky taste which will still be noticeable in future batches. Tobacco is an example of how man can develop unhealthy products from nature. However, the negative effects of tobacco can be used to advantage by brewing tobacco tea and spraying in your garden as an effective insect repellent! Under no circumstances should tobacco be left in a food container, as the tea could cause death if drunk. This happened when I was a child.

One of our neighbours grew tobacco in his garden, storing the dried leaves in his kitchen. The family kept a large variety of herbal teas and, by mistake, the tobacco was used to brew tea, with deadly consequences!

A SEPARATE FUNGUS IN CASE OF EMERGENCIES

As with all living organisms, the Kombucha fungus can develop problems which will cause it to become useless or even die. I always recommend that you have a separate fungus stored in the refrigerator to ensure the continuation of Kombucha brewing in case of an accident. The stored fungus should be taken from a healthy specimen, that had previously achieved good brewing results. The fungus should be placed in an airtight container with approx. $1^1/_4$ US cups ($^1/_4$ lit./$^1/_2$ pint) of brewed Kombucha beverage. The fermentation liquid will sour very slowly in the refrigerator and will be as sour as vinegar after 3 - 6 months. When using this fungus and vinegar to start a new batch, you will find that the fermentation process will be accelerated and the fungus will develop a new layer much more quickly.

WHAT TO DO IF A FUNGUS GETS TIRED OF REPRODUCING

It can happen to every Kombucha brewer — the fungus will grow tired or not perform so well, losing its full strength and ferment the tea slowly. While the fungus may still work and fermenting the sugar, the beverage may taste flat. You should switch to green tea. Also, it helps if a starter beverage is used from a more lively fungus to restore the tired one. *(Perko).*

KOMBUCHA, AN ALCOHOLIC BEVERAGE?

When brewing Kombucha, sugar is fermented and this produces alcohol. The amount of alcohol depends upon the temperature and amount of sugar used. With a minimum sugar content of $^1/_4$ US cup (50gms/2ozs) the beverage will have 0.1% alcohol after 14 days and approx. 0.3% alcohol after 21 days. By using larger amounts of sugar, $^1/_2$ US cup (150gms/6ozs) the beverage may contain up to 2% alcohol after 14 days. At the same time, however, the beverage will have a very sour taste and be practically undrinkable. With the recommended amount of sugar, 6tbsp(3ozs/80gms) per litre, and a fermentation time of between 5 and 10 days, it can be assumed that the alcohol will be of very small proportions. Pastor Weidinger states that Kombucha has very small amounts of alcohol, no more than 0.5%. The beverage can, therefore, also be given in sensible amounts to children and teenagers. "Recovered alcoholics do not have to fear the small amounts of alcohol." In fact replacing the alcoholic drink with Kombucha may even help people with a drinking problem.

Interesting in this context is a letter from Mrs M.T.: *"I tried Kombucha with blackcurrant and it tasted great. I no longer need that alcoholic drink to relax when I get home after work. Kombucha is better - it doesn't cause aches and pains, and it doesn't keep me awake at night."*

CAN THE FUNGUS BE EATEN?

This question emerges time and time again. I ask myself, 'Why would anyone want to eat this fungus? It is as tough as leather and slimy!' I do not find it very appetising. However, the fungus is not harmful. Only once in any literature dealing with Kombucha did I find a specific reference to eating Kombucha, from Professor Lindner, a German researcher, who observed that the slippery mass easily passes the bowel walls and aids in constipation (see p.80). The fungus itself has all the healing properties the beverage has, but in a very concentrated form. A pressed extract can be obtained from the fungus and is sometimes sold in health food stores. If the pressed extract is a concentrated medicine, why indeed should one not eat the fungus, if one finds it appetising? *[see "Kombucha used in Veterinary Medicine" p.92]*

LIFESPAN OF THE FUNGUS

If properly treated the fungus will last for some months when used for continuous fermentation. The surface constantly builds new layers while, on the underside, the layers die off. In the batch brewing method, your original fungus will last 4-5 brews while producing a succession of new offspring. If the fungus is well cared for, it is a gift which can be given from generation to generation, as has happened for thousands of years.

KOMBUCHA — PRESSED EXTRACT FOR DIABETICS OR ON TRAVELS

In the late 1920s, researchers Wischofski and Hermann produced Kombucha in a concentrated form. The idea was to achieve a pure product which would be available with a certain consistency, which was important in making comparisons for research purposes. Known as Kombucha drops or 'Kombuchal', this product was patented in Germany and available in pharmacies. The drops were manufactured by vacuum distillation of the liquid culture. Apart from alcohol and acetic acid, Kombuchal has all the components of normal Kombucha. From the records of the Internal Medicine Clinic in Prague it was determined that all tests showed results favourable with old age and arterial blockage.

The idea of pressing the Kombucha fungus came from its most famous researcher, Dr Sklenar himself. The product was named 'Drops Kombucha D1' (D1 being the homeopathic dilution degree of 1 to 10) and was also available as a Kombucha tincture. With the drops he achieved healing results which gave him a reputation that lasted long after he died. People who do not wish to miss the beneficial effects of Kombucha are advised to take Kombucha extract with them in this handy form when travelling. It also seems to help with travel sickness. Kombucha drops, or pressed extract, are highly concentrated forms of the effective parts normally found in the beverage. Manufacturers recommend the use of 15 to 20 drops, three times daily for adults, to be taken with water.

Kombucha drops or concentrated pressed extract would also be beneficial for diabetics who are concerned about the use of sugar in the fermentation *(see p.81 "Diabetes")* Kombucha is recommended for diabetics, but, because of the difficulty of determining the sugar content in a home-brewed beverage, it would be better to use a more sour-tasting, or a longer-fermented brew.

HOW TO STORE KOMBUCHA ONCE IT HAS FERMENTED

The Kombucha beverage is a living drink, with a healthy and energising effect. Fermentation occurs more quickly at a warm temperature and slows down considerably when bottled and stored too long in a refrigerator. If there is insufficient space to store all your brew, then place the bottles in a cool place e.g. the garage or an out-building. The end result will be a sour beverage that is still healthy, but not to everyone's taste. During slimming diets, this sour beverage is used frequently. Caution should be taken with bottles stored for a long period, however; they may explode due to a build-up of gas.

Commercial Kombucha beverage, available in some health shops, may occasionally have a sour taste. This would be a good indication that it has not been preserved or tampered with. It should be as nutritious as 'home brew'. It is possible that, when first tasting some commercial or over-brewed Kombucha, it may be unappealing and sour to some palates. Kombucha should taste pleasant and one should not forget that unlimited different flavours can be achieved by experimenting with various teas and herbs. Depending upon the length of the fermentation, the liquid will be either clear or cloudy. After a short period of time, the yeast will settle on the bottom. This yeast sediment can either be drunk, as with 'wheat beers', or filtered off with a cloth. Between 10% – 20% of the liquid should be used to start the next batch. The rest may be poured into bottles and sealed — taking care to leave a little gap at the top to prevent the bottles from exploding.

DOSAGE, HOW MUCH CAN ONE DRINK?

Kombucha is a very good detoxifier of the liver and, as with all good bodily cleansing systems (such as fasting), may cause some discomfort if consumed in too large quantities when beginning treatment. It is advisable to start slowly and work up the amount taken gradually, week by week. Try two tablespoons, three times a day, to begin with. In this way, discomfort can usually be avoided. If you have any serious illness or extreme weakness, you should start with an even smaller dose. As little as one teaspoon three times a day can be beneficial, gradually building up the amount.

Possible side-effects, for some people, may include headache, stomach-ache, nausea, fatigue, dizziness, mild diarrhoea, constipation, pimples, rashes and wind. However, these are usually only temporary reactions, lasting from a day to a week or so in basically healthy people. Drink extra water to counter them. People with existing disorders may experience a healing crisis if they drink too much Kombucha too soon. If this happens, seek the guidance and support of your health care practitioner. The general rule, if you have a strong reaction to the beverage, is to cut down on the amount you drink or to stop for a day or two before beginning again, drinking smaller amounts, and gradually increasing this to the recommended dosage. The beneficial effects of Kombucha are usually felt first in the weakest (or most afflicted) part of the body, then the second weakest and so on.

Pastor Weidinger has had experience of drinking Kombucha for well over fifty years including brews from such countries of origin as Formosa and Taiwan. To this question, he has made the following comments: *All natural healing remedies give life. Living organisms are being given into our living body, absorbed, processed and sorted through our blood circulation. When I drink something it is not just merely 'open wide and down the hatch'! No! Much more happens. For one thing, the beverage becomes part of myself. I react to it usually in a way quite opposite to what I expect the reaction to be. First of all Kombucha will effect our soul, our feeling. We will feel happy, relaxed, satisfied. Then our consciousness will react; the ability to absorb. We find it easier to concentrate, to make decisions. We are less likely to forget important issues. It affects the activities of our mind, our intelligence. Last but not least, our body will react. So how do we learn to drink Kombucha properly? Each individual must discover his or her own answer, as every individual acts differently, feels differently. You are an individual, with you everything functions your own way. The groundwork has been laid by our Creator. You do everything differently from other people. A lot may be similar, but not everything is the same. This fact is very important to consider when deciding how much Kombucha one should drink. After consulting with a doctor, you could drink an eighth of a litre daily after the main meal to help the digestion. This should be based on an exact time plan with rest periods.*

EFFECTS OVER A LONG PERIOD OF TIME

I believe that nothing should be used over long periods of time, as our bodies will get used to it and the required effect will not last. Animal tests have revealed that vitamin treatment is much more useful if adopted over limited periods of time, rather than continuously. It is also much more economical. I have tested a lot of herbs on myself and, of course, also Kombucha, in very large dosages to determine whether there were any negative side effects. After six weeks of drinking two litres of Kombucha daily I could not find any side effects with Kombucha. In most homes where Kombucha has been consumed over decades, people usually take an occasional break of between one and two weeks. When under stress, if there is a virus going around, or when the body indicates they return to drinking Kombucha again.

How long should one drink Kombucha? Pastor Weidinger recalls that he once received the following letter: "Through a friend I received a Kombucha culture and I have already started brewing. Is this drink for permanent consumption or should I stop in between? How can I keep the culture alive while not using it?" His reply was, *The length of any therapy depends upon the motive. If I merely want to live a healthier life, I would recommend a three weeks' programme, drinking one eighth of a litre (a wineglass) daily prior to a meal, preferably starting after a full moon and ending one week prior to another full moon. If I wish to treat specific illnesses such as bowel problems, strengthening of the liver, rheumatism, gout or improvement of blood quality, I would recommend drinking three-eighths of a litre (3 wineglasses) of Kombucha daily, over a period of between 3 - 6 months, then stopping for one month before continuing again. Permanent use is not recommended, as the body requires a break between treatments in order to make proper use of the help being offered. If one has a break between treatments, the user will feel the effect much better when starting again.*

This is called the 'Interruption Theory', which is now recommended by a number of Kombucha experts, to allow the body to consolidate a new balance.

DOES KOMBUCHA HAVE NEGATIVE SIDE-EFFECTS?

Kombucha brewed with green or black tea will enhance the effect of these teas. It should therefore be no surprise that children may not be able to sleep after a glass of Kombucha in the evening. Herbal teas need to be used in a balanced way, not over-used in Kombucha brewing, but rather interchanged with other appropriate herbs for your particular needs.

The properties of the Kombucha brew may change if the brew still tastes sweet after a certain period of time. It could happen when the culture has been overheated in transport, or heated on a stove. I heard of one person who put the brew on the stove for 'temperature controlled brewing'!

The tea (green, black, herbal tea, etc.) is the part of the Kombucha brewing which most people have difficulty understanding. In my experience negative side effects are not due to the culture itself, but can be traced back to the **additional ingredients** and the fermentation process. The herbs, green tea and black tea included, are transformed in the brewing process. The properties are amplified during fermentation and more easily assimilated and utilised by the body. It is incorrect to believe that herbs can only do you good. If used incorrectly they can have negative side-effects. *(See Part IV (p.55) for more detailed information on the use of teas and herbs).*

The amount of sugar, the time of fermentation, the average fermentation temperature and the time and temperature after the main fermentation in the refrigerator all have their effect.

Hundreds of people have contacted me asking about Kombucha brewing. There have only been a few people who have had negative effects. Here are some examples:

Constipation (one case). Black tea was used and the fermentation time was not long enough, causing the drink to be still very sweet. Black tea can cause constipation, some people use mixtures of buckthorn, elderflower and knotgrass and raspberry instead, with good results.

High blood pressure. Green tea had been used in this case. I suggested using a mixture of hawthorn and other herbs instead. This was successful. See also section on "High blood pressure (page 83)".

Candida albicans (a few cases). In these cases, too much Kombucha had been taken too fast. See also section on "Candida albicans (page 78)".

Heartburn and a burning sensation in the stomach have been reported by some people. All of them had drunk Kombucha on an empty stomach. With a pH of 3 to 3.5 Kombucha can trigger heartburn. If you do not feel well after drinking the brew on an empty stomach, then don't do it! Eat something first.

Menopause: Some women reported that they had started menstruating regularly after starting to drink Kombucha. My advice was that they should be careful not to become pregnant. In two cases the women were not very pleased with my comment and explained that they were quite happy to have stopped menstruating. They then asked what they

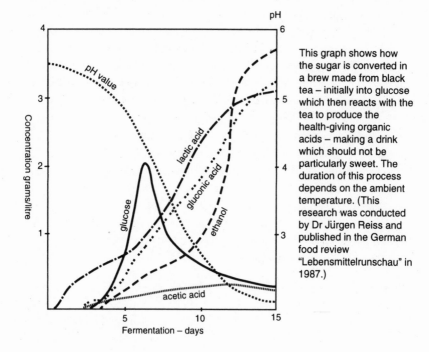

This graph shows how the sugar is converted in a brew made from black tea – initially into glucose which then reacts with the tea to produce the health-giving organic acids – making a drink which should not be particularly sweet. The duration of this process depends on the ambient temperature. (This research was conducted by Dr Jürgen Reiss and published in the German food review "Lebensmittelrunschau" in 1987.)

could do about it as they liked the positive effects of Kombucha. The only advice I was able to give was that they either enjoy their youth or stop drinking Kombucha.

Menstruation can be affected by Kombucha drinking. Some women reported that they experienced a stronger and prolonged menstruation. It is recommended that women in this situation should stop drinking Kombucha one week prior to their expected menses. They should also consider brewing Kombucha with other herbal teas.

Nose bleeding was suggested in one case after drinking Kombucha. The advice was to stop drinking the brew or at least lower the amount drunk. Some believe that Kombucha thins the blood. Those using the drugs warfarin, heparin and other drugs which thin the blood should seek advice from their doctor.

The majority of people have very good results with Kombucha, but there is always the chance that a few people will have a negative reaction, for whatever reason.

THE TRANSFORMATION OF SUGAR IN KOMBUCHA

When they read in the recipe the amount of sugar needed for brewing, a lot of health-conscious people stare in disbelief. So much unhealthy sugar? The answer is Yes! The Kombucha culture requires white sugar to ferment and stay alive (unrefined sugar does not work well). Through fermentation white sugar gets transformed into lactic acid and alcohol. With less than ¼ US cup (50gms/2ozs), the fungus will starve.

Pastor Weidinger answers the question in the following way: *Sugar on its own may cause blood disorders and depression. As a sweetener, it has been rejected by our society for years now. Many experts agree that the constant use of sugar has a negative effect on the blood quality. This may then result in poor functioning of the liver and, in turn, influence our moods. So how about the use of sugar in Kombucha? As the sugar in the Kombucha beverage is completely converted, there will be no negative effects of any such kind.*

Recently a person (can of Coke in hand) tried to make me aware of the damaging effects of the sugar in Kombucha. She had problems in believing that in every Coke she drank there were not only seven tea-spoons of sugar (unfermented!) but also 32mg of caffeine!

HONEY INSTEAD OF SUGAR?

The use of sugar continues to be a matter of concern for many health-conscious people. However, the fact is that the fungus requires sugar to survive. It ferments into 'healthy' components. With artificial sweeteners, the fungus would starve and die. The effect of honey on the fungus can also be detrimental and cause it, eventually, to die. Sugar is essential for the fungus's own digestive process. This is also the reason why the amount of sugar does not necessarily make a dif-ference to the taste, once the fermentation process is completed. The ideal amount of sugar per US quart is between ¼ to 1 US cup (2-8ozs/50-200gms per litre), depending on recipes. If less than ¼ cup (50gms/2ozs) is used, the fungus will starve; over one US cup (200gms/8ozs) means that the amount of sugar remaining after fer-mentation will be too high, which may pose a problem for diabetics as well as people with cancer or stomach and bowel illnesses.

When I conducted trials using honey some years ago, the fungus became tired and slowly died. I was surprised, however, when I met Mr

Perko from Kombucha House in Queensland, to learn that he had successfully brewed Kombucha using honey instead of sugar. Mr Perko has attempted more trials brewing Kombucha with herbs than any other person I have met. He is now making further experiments to assess the effects of various types of pollen in the honey which may have an effect on the Kombucha culture. With the mixtures of various teas that Mr Perko has tested he always used elderflower. Elderflower tea can be used to start a Kombucha fungus, it has the effect of giving the Kombucha beverage a certain champagne character, (a typical example of the many different achievable varieties of Kombucha). Mr Perko has an extraordinary feeling and great skill when brewing Kombucha.

If you do decide to try using honey for Kombucha brewing, an average temperature of 83°F (28°C) is very important. This is why experiments in tropical Queensland have been more successful. Mrs J.G. from Queensland has devised a simple method which has worked very well over a long period of time. She uses honey in every second brew, so that the individual fungus ferments one week with honey and the next week with sugar. With several containers on the go, she always has a honey brew which is mixed with the conventional brews and has produced an excellent tasting drink.

Mr Bill Corrigan from Queensland, Australia, has also written of his observations of experiments with Kombucha brewed with honey: *My own rational mind could not accept the judgment on honey, particularly when you consider that cane sugar has only become prominent in recent history and would not have been used or known in most cultures that used Kombucha. The only problems I had were in the colder months when temperatures varied. When there are uniformly warm temperatures — no problems. I always brew using green tea and some herbs separately, and allow the mixture to cool until warm **before** I add the honey. The reason for this is that no enzymes are lost to the hot liquid. The amount of honey — no less than four tablespoons of honey per litre, preferably more."*

Bill uses fresh herbs from his own garden like nettle, dandelion, yarrow and elderflower, which he adds to the brew two or three days before the end of fermentation. He stores Kombucha in bottles for two and a half months at normal temperatures under his house in tropical Queensland. He was delighted to find the taste improved with age, and was not sour.

For brewing with honey, temperature-controlled containers are necessary in cooler areas. That is the reason why my own trials did not have successful results and why publications from cooler countries say a strict NO to honey brewing.

KOMBUCHA VINEGAR

It has probably happened to a lot of Kombucha brewers that one batch has been forgotten and has continued to brew over many weeks. The result is a Kombucha vinegar which can be used like any other vinegar and can also be used to clean the fungus should it have become mouldy. Kombucha vinegar can also be added to a brew as a booster when the fungus has become tired and the resulting brew is flat.

PRESERVATION OF KOMBUCHA

If you treat your fungus properly you will have the joy and gift of receiving health and enjoyment for life. The fungus itself is a very stable living product. If you have too many individual fungi or go on vacation, the fungus may be preserved by placing it in enough of its own liquid to cover and storing it in the refrigerator. The fungus then falls into a dormant type of sleep. The container in which the fungus is stored should only be half-filled at the most, so that the fungus will still be able to breathe. It is possible to store it in this way for a period of three months without any problems.

Whether the fungus can be frozen is debatable. Some experts say there are no problems with freezing the fungus but others, such as Dr Meixner, contend that the frost crystals could be harmful to the fungus structure. Trials may have been conducted with different cultures. Personally, I do not see any reason why it should ever need to be frozen since the fungus can survive for a considerable period of time in a refrigerator.

Drying the Kombucha Fungus
Drying the fungus is another way of conserving the culture, especially for mailing. The dried fungus is very light, can be sent quite cheaply by airmail, and does not need special containers which might get damaged or leak in transit. However, dried cultures do not ferment if they are overheated or over-dried. The best drying temperature is 91°F (33°C).

Greater heat will kill the culture as will drying in direct sunlight or in a microwave oven. Dehydrators used for fruit and other foods are ideal.

Starting a brew with a dried fungus:
For the first brew using the dried fungus, it is recommended to use both tea and sugar at double strength in a smaller amount of water – 2¹/₂ US cups (¹/₂ litre/1 pint). Add commercial Kombucha, or a tblsp vinegar. A temperature of 70°-84° (23°-28°C) should be maintained for the first week. Ferment this brew for at least 15 days until it tastes very sour. Use the 'baby' from the reconstituted fungus and proceed as usual with your brewing.

Here is a suggested recipe:

Ingredients:
2¹/₂ US cups, (¹/₂ litre/1 pint) water
6 US tablespoons (80gms/3ozs) sugar
1 teaspoon (or teabag) black or green tea
If available, ¹/₂ US pint (200mls/ ¹/₃ pint) commercial Kombucha beverage *or* one tablespoon vinegar (apple cider or white)
One dried Kombucha fungus

Method: use Method on p.23.

KEFIR, YOGHURT AND SIMILAR CULTURES

I am continually being asked whether Kombucha can be compared to Kefir or ginger culture. With Kefir or yoghurt, cultures are used for the fermentation process of milk. The most popular of these three is yoghurt, which has been available for well over fifty years. Natural yoghurt (without the addition of any fruit or flavours) can easily be produced at home. The micro-organisms in yoghurt *(Lactobacillus bulgaricus and streptococcus thermophilus)* ferment milk within three hours when kept at 110°F (44°C). Fruit can then be mixed in if desired.

I am asked again and again about ginger culture used for the manufacture of ginger beer. With the brewing methods of today, ginger beer is fermented with the beer yeast called *Saccharomyces cerevisiae.*

Kefir (sometimes called Tibi) is another culture similar to

Kombucha, using water (or in some cases milk) in the process. Kefir appears originally to have come from the Caucasus. The culture consists of small white transparent pellets which, when cultivated in liquid, increase at a rapid rate. The fermentation time is approximately two to three days, the gas content increases day by day. Similar to Kombucha, Kefir requires sugar to live. Sultanas and dates are added to the fermentation process as an aid and to improve the taste. Kefir is also known as Japanese crystals and Water Kefir is a beverage generally used to improve water, and in this way, overall health. It aids the body in the detoxification process, which in turn helps stability. The recommended amount to drink per day is approximately one litre.

KOMBU-TEA

The use of the name 'Kombu' in Japan is rather confusing. Reports about the health benefits of 'Kombu-cha' ('cha' stands for tea!) from Japan are probably based on the sea-vegetable-tea Kombu (kelp = Laminaria family). 'Kombu' is harvested in salt water and has nothing to do with the tea-fermenting fungus 'Kombucha'.

I have discussed Kombucha with Japanese people who have used this term interchangeably for the salt water 'Kombu-tea', before I realised we were talking at cross purposes!

A healthy culture

Part III:
HOW DOES THE FUNGUS WORK?

KOMBUCHA FUNGUS – A BIOLOGICAL PRODUCTION CENTER

Today we consume vast quantities of preservatives, cook with microwave ovens and use other processes which destroy the living energy in our food. It is important for our digestion to consume living micro-organisms and bacteria. Without these tiny micro-biological helpers, however, we and plants would not be able to survive. For example, approximately 250 million bacteria work in one gram of good gardening soil. Wine and beer would not exist without micro-organisms. It is a miracle that these invisible helpers can adapt at all, though now they seem to be able to survive in our chemical age albeit with increasing difficulty, coping with approximately 3000 food additives in our daily diet. We should not be surprised that new illnesses and allergies are created with this bombardment of foreign products and additives.

Pastor Weidinger comments on Kombucha as follows: *The breeding does not happen through spores as with conventional yeast cultures. Kombucha is a lichen. Lichens are regarded as the oldest food and healing components of man. These remarkable plant organisms evolved some 2.5 billion years ago. It began with sea algae which were confined to the space of the ocean. When they moved onto land, they were dependent upon a mate, which they found in the fungi ('mycel'). They formed a living relationship and evolved into one unit, the lichen or lichens. Only in this form were they able to survive. The algae supplies the fungus with organic nutrients and carbohydrates. The fungus had the function of serving as a water reservoir also containing vital minerals.*

There are well over 1600 types of lichens. The manna lichen delivers food for man, the reindeer lichen supplies food for animals. Others such as Irish moss and lung lichen act as healing and medical components, or the Oak moss for the manufacture of perfumes. Lichens incorporate strength. They seem to be messengers from God which came to us from another world. Although there is a large variety of them, they all have the same basic principles, the lichen acids. These are vital components of the lichen which revitalise living matter, minerals and trace elements. Where there are lichen, life is awakened. The tea fungus Kombucha is more than just a fungus; it is a symbiosis of algae and fungi. Kombucha is a lichen. Kombucha tea is life, receives life, aids life. A fully grown Kombucha lichen builds a gelatin-like tough membrane, assuming the same diameter as the container in which it grows.

The fungus expert, Dr Meixner, writes that the important components of the Kombucha fungus are the tropical yeast *Schizosaccharomyces pombe, Saccaromycodes ludwigii and Pichia fermentans.* The most important bacteria in the symbiosis is apparently the slimy vinegar bacteria *Acetobacter xylinum,* which creates a slippery mass out of cellulose in the symbiosis. These bacterial and fungi parts are bound in the tea fungus, therefore enabling symbiosis.

Dr Helmut Golz mentions the following five active bacteria in his book Kombucha - *Healthy Beverage and Natural Remedy from the Far East.*

- *cetobacter xylinum*
- *Acetobacter xylinoides*
- *Gluconobacter bluconicum*
- *Acetobacter aceti*
- *Acetobacter pasteurianum*

and as yeasts he mentions:

- *Schizosaccharomyces pombe*
- Yeasts originating from *Apiculatus*
- *Saccharomycodes ludwigii*
- *Pichia fermentans*
- *Mycoderma*
- Yeasts originating from *Torula*

The split yeasts *(Schizosaccharomyces)* increase in contrast with conventional yeasts by cross-division. This appears to be the reason why Candida albicans sufferers can enjoy yeasty Kombucha beverage *(Gunther Frank).*

KOMBUCHA CULTURES ARE NOT IDENTICAL — IS THERE AN AUTHENTIC KOMBUCHA?

There is no exact answer to this question. Kombucha cultures have existed on this planet for thousands of years. These tea fungi were grown in conventional kitchens and have different stocks or combinations of yeast and bacteria. I think the easiest comparison can be made with apple trees. In an orchard, there are twenty apple trees which look the same. Which is the best tree? This question can only be answered at the time of harvest. Some trees will have sour apples, some sweet ones, some juicy. The continuation of the stock will only be achieved by using the best. Pastor Weidinger explains his selection process as follows:

Only the best mother animals and mother trees will produce good stock - a fact of which every grower is aware. The important factor is to reproduce something living that has its own unique qualities and cannot be exchanged with something else. The same goes for reproducing a Kombucha fungus. Only a large, solid and well-developed Kombucha fungus will be capable of being divided.

Research into the origin of the culture shows that it has been grown for over two thousand years in stone jars in unsterile farm kitchens. Thanks to its antibiotic resistance, it has retained its health and stability. Is the laboratory culture, grown hygienically for 30-40 years, still as stable? During brewing, the culture has to be kept clean, otherwise mould will build and the culture will die. In some cases the culture will still ferment, but the beverage will not be as lively and the effect will not be the same as that of a healthy culture. It is important to continue brewing with a fungus that has the best qualities. In case an 'accident' happens and the fungus does not work as it used to, it is helpful to put the fungus into 'mother tea' for a period of time to help produce the missing bacteria and yeast components.

The effect of the varying beverages can be compared only when their conditions of culture are the same, that is when they include:

● *the same type of culture*
● *the same sort of tea (black, green, herbal tea, etc.)*
● *the same amount of sugar per litre*
● *the same time of fermentation*
● *the same average fermentation temperature*
● *the same time and temperature after the main fermentation in the fridge - during which the beverage will still continue to ferment.*

Variations of the above can achieve many different tastes and effects on performance. Different combinations may cause problems for some people. It is therefore particularly important to use only the best and healthiest cultures for further brewing and multiplication. A fungus which is not working properly should be discarded. Kombucha is the tea with one thousand and one different tastes. The purchase or gift of a Kombucha culture requires trust. If you need help in obtaining a healthy culture, see *Appendix Two* (p.105).

CAN KOMBUCHA BE CONSIDERED A MIRACLE HEALING MEDIUM?

When I hear terms such as miracle healing medium, it normally makes me rather sceptical. Just as there is never only one cause for any particular illness, so it is unlikely that one general miracle medium exists. There is general agreement that Kombucha balances the digestion and helps regularity. 'We are what we eat' is an old saying. Hippocrates said: "Death sits in our bowels", and links many illnesses to problems with the digestion. According to experts, the healing of the bowel flora is the first and most important step to curing an illness. In traditional medicine, the application of Kombucha showed such positive results that words such as magic and miracle were associated with it.

LACTIC ACID FERMENTATION

The term 'lactic acid fermentation' does not sound very appetising. However, the products associated with this term are not only well known but also very popular. What would we do without wine, beer and cheese on our menu? Lactic acid fermentation plays an important part in preservation. One of the best examples of this process is sauerkraut. The famous sea captain, Captain Cook, always took sauerkraut with him on his expeditions, as this was the only way to preserve vitamins in those times. Without the important vitamin C (approx.20mg per 100gr of sauerkraut), the men on those ships would have contracted scurvy. Today salt is mostly used in sauerkraut but in the early days people used herbs and spices such as juniper berry, mustard seed and thyme, instead of salt. To avoid calcium deficiency, it was common to add eggshell flour to the sauerkraut.

As a strengthening beverage for manual workers, a vinegar beverage was mentioned three thousand years ago. "Come here, eat your bread and dunk it into your vinegar beverage" (*The Bible,* Ruth 2,14). The farmer, Boas, used these words to invite Ruth, later to be his wife, for a meal. The advantages of lactic acid fermentation were already well-known in ancient times.

Sour dough bread is the basic food source in many countries. It keeps fresh longer and is easier to digest. Particularly interesting was a conversation I had with a Maori lady. I explained to her how to brew Kombucha, how to multiply the fungus and so on. She asked if this was similar to dealing with the bread-plant *(R'ewa'n'a P'arawa)*. It was news to me that, in this part of the world, fermented bread was traditionally produced. R'ewa'n'a P'arawa was handed down from generation to generation as a religious gift. The bread was made from the roots of a fern and ensured health and strength.

Kombucha fermentation is a lactic acid fermentation process. There are two lactic acids, one being 'L(+) lactic acid' and the other 'D(-) lactic acid'. The (+) lactic acid in Kombucha detoxifies, cleanses, purifies and strengthens the body.

The metabolic by-products of the fermentation produced by the tea and sugar include Gluconic and Glucuronic Acids, Acetic acid, Carbonic acid and Vitamins B1, B2, B3, B6, B12, Folic acid and various enzymes. Usinic acid produced is antibacterial and partly antiviral. Acidic acid bacteria produced are antagonistic to streptococci, diplococci, flexner and shigella rods *(Nosredna)*.

ACID AND ALKALINE FOODS

There are two groups of acid and alkaline foods. One is made up of foods that are actually acid or alkaline and the other from foods which produce an acid or alkaline reaction. The reversing effect of the second group leads to false conclusions. For example, the lime is extremely acid with a pH reading of 1.9, but this fruit increases the alkaline content of the body. If you want to influence your body's pH, it is not so important to know the pH of any particular food you eat, but to know what reaction that food will have in your body. For more information, I recommended a study of the book: *Acid and Alkaline* by Herman Aihara. Research from Russia and Czechoslovakia is interesting in this context. Water, for the purpose of this research, is divided into "live water" and "dead water." DC current converts neutral water of about pH7 into acid water (pH 4) and alkaline water (pH 10). This water was used, according to reports, against a wide range of diseases with excellent results. The research was based on the fact that plants can thrive only in a specific pH milieu. Unwanted stinging nettles for example can be eliminated in two ways. One is to spray with chemicals, with their negative environmental side effects. The other is to alter the pH of the soil (calcium) so that the stinging nettles can no longer thrive.

There is a similarity with diseases, according to the research. Altering the pH of the blood should fight certain diseases. With cancer for example, if the blood pH is altered with activated water to 7.4 there is no 'fertile soil' for the cancer to grow in, which should prevent further growth of the tumour. As the information from Russia shows, over 500 patients with different diseases were all treated with success. It is also pointed out, however, that it is impossible to cure all diseases with this water. *(Krotov)* The manufacture of this water is relatively easy. I have met in the last months some people from Czechoslovakia who confirmed with enthusiasm the healing successes achieved with activated water. You can find detailed information about 'live' and 'dead' water in my book, *Water Medicine.*

THE pH OF KOMBUCHA

The pH level in Kombucha is of great importance and questions about the pH level are frequently asked. The interaction with the pH of food that we eat and how it affects our pH in the body is very complex. Fermented Kombucha, when it tastes like cider, has a pH reading of around 3.0. With varying tea/sugar amounts and fermentation time, the pH can vary between 2.8 and 4.0 *(For pH test strips – see pp.105 & 112).*

The pH scale goes from 0 to 14 and shows if a substance is acid or alkaline. Unpolluted rain water has a neutral pH of 7.0. Lemon juice lies at pH 2.0, strawberries at 3.3, raspberries 3.4, potatoes at 5.8 and salt about 7.0. With more awareness of pH and its adverse effect on our metabolism, many people become confused and draw the wrong conclusions. A woman from South Australia wrote me that "Kombucha cannot be good for arthritis since with a pH reading of about 3.0 it must have a negative effect". This is a common misunderstanding. A 'sour' or acid food does not necessarily create an acid milieu in the body. In fact, Kombucha can help create a pH balance in the body.

Another person, worrying about the pH levels in the body, bought Litmus paper and examined the pH of their urine every day to confirm whether the pH in the body was correct. The diet was adjusted accordingly to achieve a pH reading of the urine at 7.4. One would imagine that this person gave up drinking Kombucha after a short time as a low pH reading would probably have indicated that his body was functioning as it should — flushing out the excess acid. This would only be temporary and a greater pH balance would have been achieved because of it.

When someone talks about an acid body, what does it mean?

There are many liquids in our body with quite different and specific pH levels. In the digestive tract, for example, the saliva has a pH of 7.1; the stomach juices a pH of 1.5 and the pancreatic juices a pH of 8.8. *(Aihara)* Despite these variations in the body, the blood has a very constant pH reading of 7.4. Young people have a lower reading, but with increasing age the pH of the blood becomes more alkaline. Fluctuations of the pH in the blood could lead to severe disturbances, A blood pH of 7.7 can result in tetanic convulsions and a blood pH of 6.95 *(Fasching* suggests a level of pH 6.0) could lead to unconsciousness and death *(Aihara)*. A pH level of over 7.56 sets the stage for the development of tumors *(Fasching)*. It might be confusing, but acid food (pH of less that 7.0) has an alkaline effect in the body.

Part IV:
MAKING THE BEVERAGE MORE INTERESTING

WHICH IS THE RIGHT KIND OF TEA?

I am often asked which is the best tea to use for a Kombucha fermentation. In the Eastern countries of its origin only green tea was used. In Russia and the German-speaking countries, mainly Russian tea — black tea — is used. Tea is the most popular drink in the world — not beer or Coke as some people want to believe. Besides green and black tea, there is the semi-fermented Oolong tea. The source of all common teas on the market is the tea bush *Camellia sinensis*.

The teas known as English breakfast teas, Russian teas, etc. are black fermented teas. Some of these are blended with other teas, e.g. with other herbs like Bergamot which gives 'Earl Grey' its flavour. Most people in the Western world drink black or Oolong tea rather than green tea, probably because of the stronger flavour. Many individuals feel a stimulating effect when drinking black tea or coffee. Green tea, on the other hand, doesn't stimulate as much, but has a steady and longer-lasting effect. Caffeine stimulates the central nervous system as well as the action of the lungs and heart and promotes urine production.

In recent times green tea has frequently been mentioned as a potential anti-cancer aid. Robert E. Willner (USA), author of *The Cancer Solution* mentions "delicate pale tea" as a source of anti-carcinogenic substances. Green tea contains the chemical 'Epigallocatechin Gallate' (EGCG) which inhibits the growth of cancer and lowers cholesterol levels. "Green tea gives health and longevity" is an old Buddhist saying.

The Japanese go slightly further in utilising the healing benefits of green tea — they eat the tea leaves after drinking the tea, thus consuming the precious vitamins and trace elements contained in the leaves which are normally thrown away. Professor Kazutami Kuwano of the Kasei Gakuin Junior College, closely investigated the 'tea-food'. He came to the conclusion that the high content of vitamins A and E which are contained in green tea are not absorbed by the body since these vitamins are not water-soluble.

Calcium, iron and vitamin C are only half utilised in tea when drunk, compared with that absorbed from the tea when eaten. To make three cups of green tea one uses approximately six grams of tea leaves — a teaspoonful. This quantity of tea when eaten contains 50% of our daily needs óf vitamin E, and 20% of our vitamin A requirements. The healing effect of green tea is determined by the quality of the product. This depends on where the tea is grown, the harvest, whether vital young leaves or old ones were used and the processing procedures. Loss of quality of tea or of any other medicinal herb can also occur with poor conditions or too long storage.

HERBAL TEA REFINEMENT THROUGH KOMBUCHA FERMENTATION

The fermented liquid known as Kombucha has been used by many people for centuries as a gourmet product and healing remedy. The different taste combinations are endless. Mixing medicinal herbs with common black or green tea results in a brew containing the properties of all the teas used. According to some people the fermentation process even highlights some of the characteristics of the teas.

When brewing Kombucha, a very light tea is generally used. People who react to black tea will have this effect increased with Kombucha. Beneficial results have been achieved with herbal teas, and it is well worth experimenting with some of those listed below.

When brewing Kombucha, however, not all teas are suitable. Volatile oils can harm the fungus and even kill it. A thin film of oil can spread over the liquid and starve it of oxygen. The amount of herbal tea used should be twice that of black or green tea, i.e. approx. 2 heaped teaspoons (10 grams/^{1}/$_4$ oz) per 2^1/$_2$ US pints (1 litre/2 pints).

WHICH HERBS ARE SUITABLE FOR THE KOMBUCHA PROCESS?

Some of the most popular herbal teas are: raspberry leaves, hawthorn, valerian, stinging nettle, dandelion, elderflower, strawberry leaves, blackberry leaves, etc. If you are at the beginning of your Kombucha brewing career, I advise against using herbal teas that grow close to the ground. Start with teas such as raspberry leaves, elder or hawthorn. Ground-loving teas are susceptible to larger amounts of bacteria and germs which may be neutralised to some extent but may still cause the fungus to grow mould. Although yarrow has a large amount of volatile oils, it can still be used in a mixture of about 20%. A mixture of yarrow, dandelion, stinging nettle, elder and raspberry leaves in equal parts is a very healthy mixture, which also has an excellent taste.

One thing needs to be stressed. The quality and healing extent of natural medicine depends on many aspects: how the herbs have been processed, and how fresh they are. We also need a very healthy functioning fungus culture in order to produce a good herbal tea. With herbal teas there can be some problems. I have farmed herbs for many years and have always stressed that the best herbs are those harvested from one's own garden. All herbs, with few exceptions, need to be fresh. A general rule is that once you have your new harvest, the old belongs on the compost heap. People who wish to have good results need to know when the herbs were harvested. Roots and barks generally have a longer shelf life.

Other important factors relate to the time of the harvest and the way in which the herbs have been dried and stored. Some herbs require harvesting in the morning dew and others in the afternoon sun. Most roots should be harvested during rest periods of the plants, others when the plant is in blossom. The drying process has to be done quickly to avoid mould growing which, in turn, could harm the Kombucha fungus at a later stage. Herbs should not be dried at too high a temperature and direct sunlight should also be avoided during the drying process. People who have compromised the manufacture of a health remedy and had poor results should not then say that natural medicine is ineffective. In 1984 I produced a video, entitled "Back to Nature with Medicinal Herbs", which covers the basic requirements for obtaining the correct herbs. *(see p.107)*

Most herb books on the market today are based on the experiences of our ancestors. We have inherited our knowledge of health from a time when no pills were produced and when research and industry did not attempt to obtain active ingredients, extract them and sell them in marketable form. Knowledge of healing has been built on successful experiences of working with nature. When the balance of nature is disturbed, we have to have experiences spread over long periods of time before we can be certain of the effects. Will new experiences with an unbalanced nature be a successful experience? Time will show the effects on us all.

HERB COMBINING

There are many mixtures one can develop with different teas for different purposes, for instance to aid slimming, sleeping or the treatment of asthma and bladder complaints. It is advisable to alter the mix after a few weeks to give the body a new stimulant for better effects as well as to avoid negative side effects. When making your own herbal mixtures you should consider the following points:

Is it a mixture for general wellbeing?
In this case the best taste is important. Green or black tea can be mixed with all kinds of teas, for refreshment and taste alike: blackcurrant, rosehip, raspberry (fruit and leaves), lemon, tropical fruit teas, and so on.

Is the tea beneficial for you?
When you are suffering from constipation, for example, peppermint is not the right tea to use. Green and black tea can affect your blood pressure. If you don't feel well after drinking a tea — use a different mixture next time!

Is it a mixture to treat a specific complaint?
All the teas used in the mixtures should be beneficial for the illness being treated. The number in the brackets after the herbs stand for the proportional amounts used: (1) = one part (eg 1 teaspoon). The following are only a few examples from my own mixtures:

ASTHMA: *Mixture A:* St John's wort (1), St Benedict's root (1), valerian root (1), angelica root (1), melissa (2).
Mixture B: Thymis (1), St Benedict's root (1), elderflower (1-2), marshmallow root (1), melissa (1).
Mixture C: Mullein (1), St Benedict's root (1), ribwort plantain (1).

BLADDER COMPLAINTS: Horsetail (1), calendula (1), prostawort (1).

INSOMNIA: Hops (1), St John's wort (1-2), valerian root (1-2), melissa (1), passion flower (1-2).

SLIMMING: Lady's mantle (1), angelica seed or root (1), calendula (1-2), corn silk (2), St Benedict's root (1-2), buckthorn bark (1)

SOME MEDICINAL HERBAL TEAS AND THEIR COMMON USES

For Kombucha brewing, the recommended proportion of ordinary tea with other herbs is in brackets; for example (1-4) means that black or green tea should only be used as one part out of four: i.e. one teaspoon black or green tea with three teaspoons of the herb listed. (When mixing roots or bark with leaves and flowers, the specific weight has to be considered as well as the time the tea is left in the water before straining.)

AGRIMONY *(Agrimonia eupatoria):* A bitter tonic (strengthening). Astringent to the gastro-intestinal tract. Useful in the treatment of diarrhoea. A good remedy for liver complaints. Effective for coughs, sore throat, bronchitis (gargle); strengthens the immune response. (1-3)

ANGELICA ROOT SEED *(Archangelica)* is one of the main general tonics: stimulant, stomachic, carminative. (1-4)

BARBERRY *(Berberis vulgaris):* A good remedy for stomach, liver and kidney disorders, chronic constipation and jaundice - has diuretic properties. (1-4)

BEDSTRAW *(Gallium verum)* cleanses kidneys, liver, pancreas and spleen; supports the lymphatic system and the treatment of tumours. (1-3)

BILBERRY LEAVES *(Vaccinum Myrtillus)* must be taken with great caution and for a short period of time only, followed by an equally long break, also used as infusion. (1-3)

BUCKTHORN *(Rhamnus Frangulata):* A mild laxative, cathartic. Can trigger menstruation. (1-4)

CALENDULA *(Calendula officonalis):* To tonify blood vessels as in varicose veins and for infections especially in the lymphatic system. Regulates menstruation; blood cleansing and a mild laxative. (1-3)

CHAMOMILE *(Matricaria chamomilla):* Tonic carminative and sedative. Useful in neuralgic pain, infantile convulsions, stomach and gastro-intestinal disorders. (1-4)

COLTSFOOT *(Tussilago farfara):* Expectorant for cases of coughs and asthma. (1-3)

DANDELION *(Taraxacum officinalis):* For the stimulation and function of liver and gall bladder. It is used in all digestive disorders and in rheumatism and gout. (1-3)

ELDERFLOWER *(Sambucus nigra):* A splendid spring medicine, blood purifier, gentle laxative. (1-1)

FENNEL SEED *(Foeniculum vulgare)* is often used for babies suffering from dyspepsia, diarrhoea and flatulence. Mixed with other appropriate herbs to help relieve complaints of the respiratory and digestive tracts. Used for bronchial catarrhs and other severe respiratory infections and colic. (1-6)

GENTIAN *(Gentiana lutea):* A time-honoured tonic, promotes digestion. (1-5)

GOLDEN ROD *(Solidago virgaurea)* is very beneficial for the kidneys and urinary tract. An excellent diuretic for chronic infections of the kidneys, as it increases the flow of urine. Stimulates the blood supply to the kidneys and veins of the legs. (1-2)

HAWTHORN *(Crataegus oxyacantha)* is the cardiac tonic. It stimulates the supply of oxygen to the heart muscle. Regulates blood pressure and heart action. (1-2)

HORSETAIL *(Equisetum arvense):* For all disturbances of the urinary tract, also for bed-wetting. (1-2)

MARSHMALLOW ROOT *(Althaea officinalis):* For inflammation of the whole digestive tract and of the respiratory organs etc. (1-3)

MOTHERWORT *(Leonurus Cardiaca):* An antispasmodic heart tonic for anxiety; allaying nervous irritability due to overactivity of the thyroid glands. Excellent tonic for female complaints and nursing mothers. (1-3)

MULLEIN *(Verbascum densiflorum)* is excellent for most lung complaints including coughs, bronchitis, hoarseness of the throat. Relieves irritation of the respiratory mucus membrane. It has a cleansing and expectorating effect on the mucus in the respiratory tract. (1-2)

OAK BARK *(Quercus robur):* For haemorrhoids, leg ulcers, chilblains and foot perspiration. (1-3)

PEPPERMINT *(Mentha pip.)* is best known for curing stomach and intestinal complaints. It is slightly disinfectant, spasmolytic, carminative, promotes the production and flow of bile - helps flatulence, vomiting, soothes nausea. Indications include stomach upsets, poor digestion, stomach cramps, diarrhoea, nervousness, restlessness, colic. (1-5)

PROSTAWORT *(Epilobium roseum):* The best herb for all kinds of prostate, bladder and kidney disorders. (1-2)

RASPBERRY LEAVES *(Rubus idaeus)* have been useful with diarrhoea and stomatitis, also been used as a mouthwash. (1-1)

RIBWORT PLANTAIN *(Plantago lanceolata):* Effective in the treatment of infections of the respiratory tract; catarrh and removes secretions from the bronchial tubes. Tonifying and stabilising for tissue because of its high content of silica. Has been used as a remedy for tuberculosis. External use for wounds. (1-2)

SAGE *(Salvia officinalis)* is a blood cleanser, suppresses perspiration. In malfunction of the thyroid glands or during menopause it assists poor, irregular and painful menstruation. Gargles are the best remedies for laryngitis, tonsillitis, ulcerated and sore throats. (1-4)

SHEPHERD'S PURSE *(Capsella bursa pastoris)* is a very important homeostatic plant, especially uterine haemorrhages, menstrual difficulties, kidneys, nose, mouth, bladder and respiratory tract. Regulates high or low blood pressure. (1-3)

SPEEDWELL *(Veronica officinalis)* is a blood cleanser, regulating menstruation, perspiration and mucous discharges. Tonic and soothing effect on nerves.

ST.JOHN'S WORT *(Hypericum perforatum):* Useful in nervous disorders, depression and hormonal imbalance. External oil applications are effective in the treatment of wounds, burns, frostbite, rheumatism and backaches. (1-4)

ST.BENEDICTS *(Geum urbanum):* For febrile conditions, asthma and obesity. It stimulates the digestion and strengthens the body. (1-3)

STINGING NETTLE *(Urtica dioica):* For urinary and skin disorders, as well as for all rheumatic complaints - to reduce excess acidity. (1-3)

VALERIAN ROOT *(Valeriana officinalis)* is very popular for use in nervous conditions and insomnia as a calmative. (1-5)

WHITE DEAD NETTLE *(Lamium album)* is effective in menstrual difficulties and vaginal discharge, external and internal applications. A uterine tonic also for the treatment of the urinary tract. (1-2)

WILLOW BARK *(Salix alba):* Reduces fever and alleviates pain. It is a very effective remedy in the treatment of rheumatism, inflammations, internal bleeding and makes a good diuretic. It can be used as a gargle for tonsils and externally for wounds and burns. (1-3)

YARROW *(Achillea millefolium):* For bleeding, haemorrhoids, varicose and menstrual disorders. It improves the circulation and is antispasmodic. (1-4)

DO HERBS HAVE NEGATIVE SIDE EFFECTS?

Where there is light there is also shadow. The amount of the dose can make the difference between a medicine and a poison. High doses of medicinal herbs can have negative effects on the digestive system, the kidneys and the liver. Powerful herbal medicine should only be used if there is a clear indication and need for this particular herb or herbal mixture and only with guidance from a qualified herbal practitioner. If you are healthy, enjoy harmless herbal teas to maintain health, and for their pleasant taste. The amount of medicinal herbs you ingest in a given period of time is also important.

Swedish bitters are a good example. This mixture of many strong medicinal herbs is a potent medicine with a broad healing spectrum. Some irresponsible literature recommends taking this medicine as a daily preventative, but it can have negative side effects when taken over a long period. I use the original 'Swedish bitters' around 5 or 10 times a year and find it a great medicine, but I must warn you not to use it daily without expert advice.

There is no doubt that it is better to prevent illness than to cure it. The problem starts here for many people. From fear of illness many people use strong natural medicine in an attempt to ward off possible disorders. This overlooks the fact that strong natural medicine can also have negative side effects when taken as a preventative in high doses and/or over long periods of time.

A typical example is comfrey *(Symphytum officinale)*. This very strong natural remedy to heal bone breaks, varicose ulcers, sprains and wounds should only be used externally as a herb. Internal use as a preventative can be harmful. Chamomile and peppermint can both have negative side effects if consumed inappropriately or in large amounts. Only recently I was asked by a woman what could be done for constipation. During a discussion, it turned out that she was consuming chamomile and peppermint tea daily. She stopped drinking the teas and the problem was solved.

Illness can be prevented by a balanced life. Food is one of the three main components needed for healthy living. As a known aid to digestion, Kombucha can become an important part of this. In this respect, Kombucha acts as a general strengthening remedy by enhancing the body's defences. Kombucha is therefore an important part in a preventative diet.

Stimulating herbs should be taken in the morning and calming herbs in the evening. For example, green, black and rosemary teas in the morning, St John's wort, valerian and chamomile teas in the evening. For more details on the negative effects of herbs read: *Herbal Remedies — Harmful and Beneficial Effects* by Prof. S. Teleology and Dr. A. S. Czechowicz.

KOMBUCHA — COMMERCIAL PRODUCTS

Kombucha beverages are already commercially available. Some people don't want to brew the liquid themselves and therefore prefer to buy it as a finished product. To find a good commercial beverage is not always easy. With a commercially made Kombucha drink we are dealing with a fully active food beverage. The fermentation process is still working and the taste may change daily. The Pool of Life Ltd in the U.K. have found a method that successfully produces a delicious, refreshing and naturally effervescent Kombucha beverage equal to a fine light wine.

It is possible that bottles purchased commercially will have different tastes, even though they have identical labels. Depending upon how long it has been stored and at what temperature, the taste may change. People who purchase a bottle to try once can be put off forever if their first taste is of a sour beverage which has been stored at too high a temperature. They will never discover that Kombucha is a delicious product. A change of taste, due to storage, however, is a sign that no chemical preservatives have been used and that the beverage is a complete living food product.

Commercial beverages are almost always manufactured with green or black tea. There are quite a number of manufacturers in Europe, some with enormous output, producing one standard beverage. In Germany, Kombucha pressed extract is made. I am aware of one manufacturer in Australia who incorporates herbal teas with a mixture of green tea. These herbal teas give the beverage an increased shelf life without destroying the micro-biological balance; at the same time the healing effect of the herbs is also incorporated into the beverage. This unique method has been developed with finesse after years of experimenting and experience at Kombucha House in Queensland, Australia by Mr Jose Perko. *(see Appendix Two)*

A hairdresser in Queensland, Australia, had so may good reports from clients using Kombucha that she developed a special Kombucha shampoo. *"Normally hair conditioner is applied after shampooing, but, when using Kombucha shampoo, the hair is silky and soft without a conditioner"*, said Karen from Herbacraft Gypsy Rose (international award-winning hairdressers). She sent me comments from some of her clients: *"Their hair is growing back where previously they were bald, even though they are not drinking the tea but only using the shampoo."* *"Her hair regained its natural colour. She began using the shampoo and drinking the tea, although the colour change began before she actually started drinking the tea. Initially she was using only the shampoo, which means the Kombucha mushroom works by external application to solve an internal problem."* *"Psoriasis, and an itchy, dry, flaky scalp, have incredibly all gone."* The other ingredients which Karen mixes into the shampoo are a business secret.

HOME MADE: THE MOST IMPORTANT STEP FOR THE HEALING PROCESS

Natural healing is not the symptom-cure of an illness with the use of medicine as a factor for becoming healthy. Natural healing is healing yourself. It takes place when a human being becomes a part of nature and heals himself in harmony with nature, applying various natural remedies. When Pastor Weidinger kindly agreed that I could quote from his various publications, he made the following interesting remark: *What I would like to especially highlight is the fact that I do not regard Kombucha as a trade product, but a home-made one. I support self-made Kombucha instead of commercial processing, because I*

regard the fact that someone who makes the effort to obtain and use
information for themselves is already on the way to healing.

KOMBUCHA - MIXED BEVERAGES

Kombucha can be mixed successfully with other beverages. Children, for example, love it with apple juice. Kombucha fermented with raspberry leaves tastes very similar to apple juice after three to four days and many children do not even notice that they are drinking Kombucha.

Adding raspberry or blackcurrant syrup to Kombucha is also very popular with little ones. Lemon is popular with young and old alike. Some people are convinced that whisky makes an ideal mixture! When Kombucha became popular in central Europe, people organised parties where the only beverages available were Kombucha mixtures of various types.

The most popular Kombucha drinks are the ones in which different tea mixtures are used, according to taste. Some people add fruits to the brew two days before the end of fermentation. This not only improves the flavour, it also adds vitamins and all the other benefits of the particular fruit or vegetable in the drink. When a seasonal fruit is used, it should be thoroughly cleaned and blended in a kitchen mixer. A strong fermentation is triggered shortly after the fruits have been added, the result is delicious. Have you ever tried raspberry, mango or strawberry Kombucha? Your guests, young and old, will like it.

KOMBUCHA CHAMPAGNE

When I visited Mr Perko recently in Queensland, I was able to try many different types of Kombucha. The experience was similar to wine-tasting in Austria or Italy. The last glass I was served was a champagne, a very fine product which I assumed was an expensive treat. I was very surprised when I was told that this also was a Kombucha beverage. Even though the exact recipe has not been published, Mr Perko revealed that he used sugar, honey, elderflowers, mulberries and a small amount of herbs. I am grateful that I was allowed to test this and that I was given a bottle of this magnificent liquid. Perhaps one day the recipe will be made public.

Part V:
THE THERAPEUTIC VALUE
OF KOMBUCHA

MODERN MEDICINE AND KOMBUCHA

Kombucha results can't all be fully explained scientifically, but experience and anecdotal evidence have made an overwhelming case for its efficacy over many centuries. In our technical age we try to explain everything scientifically. Cases in which it is not possible to do so have no value, according to the medical and scientific establishments. "Where is the scientific proof?", is a common question when people discuss Kombucha and similar subjects.

When I look at my books on agriculture which are three decades old, I cannot help but question the value of scientific facts. For now we have scientific confirmation that these agricultural books are wrong! Powerful drug companies are not interested in conducting scientific trials with Kombucha, as they can't make any money from it, because anyone can produce this delicious low-cost medicine at home. Kombucha works only in its natural balanced condition. The modern way of extracting the active ingredient in order to produce a marketable medicine is not suitable for Kumbucha.

Scientific trials with Kombucha, using different teas have revealed comparable amounts of the following important organic acids:

Tea	Lactic acid	Glucon acid	Acetic acid	Ethanol
Black tea	2.94%	2.52%	0.08%	1.07%
Peppermint tea	0.14%	0.04%	0.01%	0.01%
Linden flower	0.007%	0.06%	0.30%	0.04%

From the German magazine 'Lebendsmittelrundschau' (1987)

The lactic and glucon acids are the two most important components scientists have concentrated upon. There are numerous scientific reports: Dr S. Hermann and his team, conducting trials in Prague, came to the conclusion that the glucon acid in Kombucha could dissolve gallstones. Dr Wiesner's research (see below) also showed that Kombucha compared favourably with an Interferon product in gallstone treatment.

In his report, Dr Wiesner suggested that Kombucha may stimulate the body's resistance to illness, thus assisting the healing process. It may further be regarded as an important nutrient in its ability to slow a virus attack without having side effects. It should also be mentioned that Dr Reinhold Wiesner developed a blood test which electronically monitored energy levels in the blood in conjunction with a medicine being used. In the clinic at Omsk in Russia the following Kombucha trials were conducted:

● Dysentery in infants was treated with Kombucha and, after only a few days, it was recorded that the infants were regaining weight. The diarrhoea lessened. After one week of treatment, no further dysentery bacteria could be found in the faeces.

● High blood pressure and arteriosclerosis were clearly eased following 2-3 weeks treatment with Kombucha. Cholesterol levels sank remarkably at the same time.

● Inflammation of the tonsils was treated with remarkable success. Patients used Kombucha tea to rinse their mouths ten times a day; the liquid was held in the mouth for fifteen minutes each time. In addition their sinus problems improved.

A COMPARISON BETWEEN INTERFERON AND KOMBUCHA

In 1987, the physician and biologist Dr Reinhold Wiesner from Bremen, Germany, published the results of a study dealing with the effects of Kombucha. He tested the effects of the beverage on 246 patients in comparison to an Interferon product synthesised from the anti-viral agent that our bodies produce naturally. The patients had illnesses such as kidney disorder, inflammation of the liver, rheumatism, multiple sclerosis and asthma. Kombucha achieved better results than Interferon (203 compared with 183) for asthma, whereas with the other illnesses, the results were only marginally below those of Interferon: rheumatism 92% as effective, kidney disorders 89%, inflammation of the liver 81%, and multiple sclerosis 80%.

CAN KOMBUCHA HEAL CANCER?

Kombucha cannot be regarded as a single healing remedy. Within biological cancer therapy, however, it is well thought of in the German-speaking countries. Cancer experts, such as Dr Johannes Kuhl, Dr Rudolf Sklenar and Dr Veronika Carstens have had remarkable results in treating cancer where Kombucha significantly helps in revitalising the bowel flora of patients. Dr Reinhold Wiesner, a physician/biologist from Schwanemunde in Northern Germany, published a very interesting research paper. An anti-viral product was tested in comparison to Kombucha. Results showed that Kombucha's effectiveness was just marginally below that achieved by the other product. Details of the type of tea used were not given. Generally, however, black tea is used in Germany. In Asia and Japan, green tea or Kombu tea which is regarded as beneficial for health is more commonly used. For cancer treatment, pawpaw leaves and/or calendula *(calendula officinalis)* are currently being used.

Mrs R.E.H wrote me the following letter: *My 58 year old sister who is a smoker, two years after a right mastectomy, has now got secondary cancer in her ribs, skull and a growth in the lung. Chemotherapy was prescribed. Two weeks after the first chemo, she suffered complete hair loss and excruciating pain in her ribs, requiring large doses of morphine painkiller. She started drinking Manchurian Tea (Kombucha) about the time of the hair loss. Within two days, she discontinued the morphine as the pain had disappeared. Insomnia was no more. The hair on her head started to grow again and didn't fall out while chemotherapy was continuing, four sessions in all — four weeks apart. She could still work full time, except for one day following each chemo session. The bone cancer medication was supposed to swell her up, and weight gain was supposed to be any-. thing up to 5-6 stone! But she gained no weight, although her appetite was hearty. The growth in the lung shrank, and the specialist said to her "whatever you are doing, keep doing it". Of twenty-nine fellow chemotherapy patients, she is the only one still maintaining her job. All the others had to leave their place of employment, because they were too ill. She didn't experience the side effects of chemotherapy to such a severe degree as the others did. The scanning of bone cancer has not been done again so far.*

I received this letter also: *My father was diagnosed with lung cancer. He had a tumor in one lung the size of a small orange. His doctor advised him that the cancer was incurable, but that treatment would give him a better quality of life for the time he had left. His brother gave him a Kombucha which he started drinking immediately — 2 glasses a day. Four months later an X-ray showed no trace of cancer. We, the family, have no doubt that the Kombucha was responsible for his recovery.*

PAWPAW AS CANCER MEDICINE?

Three years ago pawpaw was only known to me as a delicious tropical fruit which, when added to lemon juice, had a positive effect on the digestion. I first heard of its medicinal qualities from a woman writing to the German newspaper in Australia *Die Woche,* describing how pawpaw had cured her cancer. Mrs H had lost any hope of being healed, when she was made aware that pawpaw leaves held a potential for healing.

A more detailed report in the *Weekend Bulletin* (Gold Coast, Australia) deals with the healing of Mrs K's cancer of the bladder. Mrs K (a 74-year-old woman) had an operation, but the cancer could not be completely removed and she was supposed to have further treatment in Brisbane. Over a period of three months she used pawpaw leaves. When she ran out of them, she used the skin of the fruit which she boiled. When she went back to the doctor for a further check-up after that period, the cancer had been healed. A check-up four months later confirmed this and Mrs K, who now feels completely whole, is quite convinced her cure was due to the pawpaw.

On the other side of the world the American scientist, Dr Jerry McLaughlin of the University of Purdue, has also used pawpaw in the fight against cancer. According to him, he has found a chemical component in the pawpaw tree that is *"one million times stronger than the strongest anti-cancer medicine"*. There are many reports that cancer sufferers have been healed by drinking pawpaw leaf concentrate.

A Kombucha-fermented pawpaw leaf extract is available in some health shops and Kombucha concentrate is also now manufactured in Australia *(see Appendix Two for suppliers).* I have found a number of practitioners who have used pawpaw successfully but who are reluctant to prejudice their work by saying so, for fear of legal implications. The first reports of pawpaw's healing abilities were made in 1978 and resulted in a demand for and subsequent scarcity of pawpaw leaves.

The *Gold Coast Bulletin* has published several reports on pawpaw's reputed medicinal qualities, including this one:

PAWPAW CANCER PLEA BEARS FRUIT. Gold Coast gardeners have responded to an appeal by cancer victims desperate to find sup-

plies of pawpaw leaves. And the Gold Coast man who, 14 years ago, first exposed the leaves as a possible cure for cancer has been tracked down to a Labrador nursing home. The story of how Stan Sheldon cured himself of cancer by drinking the boiled extract of pawpaw leaves was first told in the Gold Coast Bulletin in 1978. Now research in the United States has given scientific support to his claim, isolating a chemical compound in the pawpaw tree which is reported to be a million times stronger than the strongest anti-cancer drug.

Mr Sheldon, 88, says the discovery does not surprise him. "I was dying from cancer in both lungs when it was suggested to me as an old Aboriginal remedy", he said. "I tried it for two months and then I was required to have a chest x-ray during those compulsory TB checks they used to have. They told me both lungs were clear. I told my specialists and they didn't believe me until they had carried out their own tests. Then they scratched their heads and recommended I carry on drinking the extract I boiled out of the pawpaw leaves." That was in 1962. The cancer never recurred. Since then, Mr Sheldon has passed the recipe on to other cancer victims. "Sixteen of them were cured", he said. Mr Sheldon's recipe involves boiling and simmering fresh pawpaw leaves and stems in a pan for two hours before draining and bottling the extract. He said the mixture could be kept in a refrigerator, though it may ferment after three or four days. He said there were times when he found it difficult to obtain supplies of pawpaw leaves. "Not everyone likes you removing them. They are afraid you will ruin the tree", he said. But Gold Coast pawpaw growers have responded generously to a plea for help last week by cancer victims desperate to find a cancer remedy.

Home gardeners, a nursery operator and a tourist attraction owner have offered leaves from their pawpaw trees to people seeking a supply. One man, Vern Forrest of Burleigh Heads is a veritable Johnny Pawpaw seed. He has been growing pawpaws and giving away the leaves to cancer victims ever since he read the Bulletin's original 1978 story about Mr Sheldon. "I have no doubt that it works", he said. "I know of people walking around now who should have been dead according to their original cancer diagnosis. But the pawpaw treatment helped them to beat the cancer." Betty Ellingsworth, who manages a Nerang nursery, said cancer victims were welcome to the leaves from the nursery's four pawpaw trees. And Pat Washington has offered leaves from trees growing at her Tallai Hills Home.

SUPPORT - Bob Brinsmead, who operates the Avocadoland tourist attraction on the Tweed, also has pawpaw trees which he will make accessible to cancer victims. "I am not surprised by the American research", he said. "There is a pawpaw ointment made in Brisbane which has been very successful in treating skin lesions. And I once painted the sap from a pawpaw stem directly on to a worrying growth behind my ear. It got rid of it." Dr Jerry McLaughlin, the Purdue University scientist who discovered the anti-cancer substance in pawpaw, has become a victim of publicity. Attempts to contact him at the university's pharmacy school in Indiana are met by a taped message in which he says it is "impossible for him to respond to all the inquiries that have followed the reports of his research, which is "still in the stage of animal testing."

Mention should also be made that the skin of the pawpaw fruit can help with sores. (Warning — do not use pawpaw products if you are pregnant.)

The Most Common Recipe for Pawpaw Leaf Juice

Take seven fresh, healthy medium-sized pawpaw leaves, and wash thoroughly. Cut them up like cabbage and put them in a saucepan with 10 US cups (2 litres/4 pints) of water.

Bring to the boil and simmer (without a lid) until the water is reduced by half.

Strain and bottle in glass containers. The concentrate will keep in the refrigerator for 3 - 4 days. If it becomes cloudy it should not be used.

For Kombucha brewing the paw-paw concentrate is mixed with green or other medicinal teas. Kombucha fermentation preserves the properties of the pawpaw leaves naturally.

The recommended dose is 3tblspns (50ml/2fl.ozs) three times a day.

Pawpaw concentrate may be added to other fruit juices.

If it is fermented with other teas the dose is proportional; for example, for 5 US cups (1 litre/2 pints) pawpaw concentrate fermented with 5 US cups (1 litre/2 pints) of green tea, the dose will be 6 US tblspns (100ml/4fl.oz) three times a day.

Some people use semi-ripe fruit (with seeds and skin) which is said to have the same healing effect as the pawpaw leaves. This can be put in the kitchen blender and mixed with other fruits.

It is believed that the natives of the South Sea islands in the Pacific used pawpaw as a means of birth control. Only recently a researcher in Papua New Guinea told me that, depending on the amount taken, using the fruit, including the skin and kernels, can produce an abortion. In some cases, women who take high doses are unable to bear children. Recent British research showed that pregnant women may abort within a week if they eat one pawpaw fruit a day.

EARLY DETECTION OF CANCER INCREASES HEALING SUCCESS

The earlier cancer can be diagnosed, the better the chance of healing. Unfortunately, with the present standard of medical diagnosis, cancer is often already in an advanced state when discovered. According to Dr Sklenar, illnesses can be detected a long time prior to the acute stage of cancer. Dr Sklenar developed an early detection blood test with which he claims to be able to detect cancer through his 'blood colour' method. Dr Sklenar also found iris diagnosis a valuable help in early detection where the presence of brown to black sediments showed that the colon was not functioning well. Patients in a pre-cancer state often have digestive problems. Even though diagnosis of the iris is only successful in 80% of cases, it is still a precious tool in the early detection of the disease.

Other uncommon diagnosing methods also appear to be useful tools in the early detection of cancer. I felt that one of my close friends had an acute problem of the liver, so I consulted a practitioner without telling her of my concern; when I described the person to the practitioner, his conclusion was that my friend might have cancer of the liver. Tests conducted in famous hospitals in Germany and Switzerland had not confirmed cancer. The patient had had frequent check-ups for other illnesses in hospitals. Only $3^1/2$ years after the health practitioner had suggested cancer of the liver, my friend was diagnosed with just that, which resulted in death three months later. The question remains whether the patient might have been able to alter the outcome with more awareness and sensitivity by the medical establishment to the early warning signs of cancer. Other advice would have been to suggest changing her diet, stopping smoking and altering her bed position to combat negative energies running through her home.

KOMBUCHA, HEALING REMEDY FOR:

AIDS

Two of the benefits of drinking Kombucha most often mentioned are the energy it releases and its immune system boosting effects. I often thought that Kombucha might benefit AIDS sufferers but, initially, did not have any reports of this ancient remedy being used to treat the modern disease.

Recently, however, there have been widespread reports that Kombucha use in HIV+ communities has grown quickly. It is being said that the Kombucha tonic can reverse the symptoms of AIDS. In the USA, which has an estimated one million HIV+ infected people, the news of a possible miracle cure is spreading rapidly. In some publications it is being claimed that components in Kombucha partly "inactivate some viruses"; that high p24 antigen levels can be reversed, a rise in T-cells achieved and weight increases of 5-7kg recorded. All these are measurements of improvement. Long term AIDS survivors also report that Kombucha has a positive result.

The Kombucha culture is being passed on to friends as a remedy for AIDS, which is popular because of its palatability, its apparent safety and low cost to further commend it.

"Kombucha might have remained within the confines of the holistic community were it not for the August 1994 issue of 'Positive Living', a newsletter published by AIDS Project Los Angeles (ALPA). Under a headline claiming 'Reborn on the 4th of July' was the photo and story of Joe Lustig, a Long Beach man who has AIDS. As that month's subject of a regular profile on long-term AIDS survivors, Lustig told Paul Serchia, the newsletter's editor, of his 'startling breakthrough' on Independence Day.

'I spent more of 1993 in bed', Lustig says, 'I could hardly get to the bathroom without a walker'.

"Lustig first tested positive in 1985 and was diagnosed with AIDS in 1990, after coming down with pneumocytis carinii pneumonia (PCP). He developed a litany of increasingly dangerous infections: meningitis, encephalitis, pancreatitis, all complicated by dementia. 'I was in such a fog I couldn't pay my bills. I had this constant underlying feeling of tiredness and sickness. A very heavy toxic feeling. It was there every day when I got up, and every night when I went to bed.'

"After less than one week on Kombucha, however, Lustig says he felt better. 'I have to emphasise that my life was not worth living any more. Then I woke up one morning and my first thought was that I had died and gone to heaven. Everything felt better.

"According to Lustig, he got out of bed and did something he had not done in four years. He went for a run. He also insists that no other changes in his medications took place that might have accounted for the change.

"Lustig rides his bike three to five miles a day now, reports increased energy, longer waking hours and better cognitive function. He has also given up one of his medications. 'I haven't taken Nizoral (a drug that fights thrush) for ten weeks', *he says, the length of time he has been drinking the tonic."* (New Age Journal, November 1994)

Mrs R.D. wrote: *"I was introduced to Kombucha by an American friend who brought a 'mother' fungus from the U.S.A.. This person has been HIV+ for the past eleven years. He was quite excited about Kombucha and has been drinking it for nine months now, mainly to keep his T-cell count adequate. He is healthy and robust looking, maintaining a positive attitude, moderate lifestyle and good diet. My friend says that since he has been drinking the tea fungus, he feels more energetic and incredibly well. The drinking of Kombucha tea has swept the world of HIV diagnosed people and AIDS sufferers in the USA over the last year or so, as it seems to be a very good natural aid for them.*

Another correspondent *writes: "I started drinking Kombucha about three months ago and find it most helpful. About five years ago I was diagnosed HIV+. My T4-cell count then was 660 and had been dropping over the years to 390 before starting to drink Kombucha. Within 2 months my count was up to 570; it has not been that high for nearly 2^1/2 years. I also have found it helpful with my arthritis and sinus problems."*

ANAEMIA

From a letter I received: *"After being given a Kombucha fungus by a friend in the beginning of January, I have been taking the tea every day. Since 1977 I have had pernicious anaemia requiring monthly injections of vitamin B12. Before this condition was detected my legs started to get numb, from the soles up to well over the knees. After a couple of injections of B12 my legs went back to normal except for the soles of my feet - they remained feeling like 'wood' - no feeling in them at all, and I had a rash around my ankles due to some allergic reaction to the injection. For 18 years the doctor said I should not expect any improvement as the nerve ends were probably too damaged so I forgot about it.*

"Now one night this week I woke up because of a stinging pain in my toes, and in the morning when I got up, I suddenly noticed that I could feel the structure of the carpet under my feet, very remarkable, as I had never felt that before. And now, during the week, the soles of my feet went back to normal with no more numb feeling at all. Also the rash has disappeared."

ARTHRITIS

There have been many reports of healing and improvements to arthritis. However, it is not clear how far advanced the state of the illness had been with individual patients. It is certainly important that arthritis should be treated as early as possible and that changes in the diet of the sufferer should also be considered.

Mrs.T. wrote: *"Kombucha has helped my arthritis, my writing hand is now no longer in pain"*.

Mr W.R. wrote: *"In April 1993 I went on a trip to Tibet from which I returned after suffering a severe bout of altitude sickness. About three months later I contracted rheumatoid arthritis. I was eventually placed in the care of a rheumatologist who put me on severe medication to control my problem. In June/July 1994 I was introduced to Kombucha which I have been taking ever since. On my visit to the specialist in December, after blood tests, he said I was clear of arthritis. For the last three months I have not taken any medicine other than primrose oil once a day and I feel great."*

And these comments from letters: *"....pain relief from arthritis in shoulders and neck; ....I have found that within the first week of drinking this beverage, the arthritis pain I was experiencing in my finger has left me and has never returned; ......Since starting to drink Kombucha tea in December 1994, my osteoarthitis has improved 50% at least and I can enjoy life again. Many times I had thought of ending it due to the pain....Diverticultis, ear infections, bowel problems have all improved greatly. I can exercise now and my joints don't swell."*

ASTHMA

Asthma is a disease spreading at an alarming rate, triggered by many differing environmental factors and stresses. Kombucha has a positive effect upon digestion by detoxifying the body which eases the asthma. Generally speaking, asthma sufferers reported a considerable improvement when using Kombucha. Several herbal mixtures have also had a positive effect with asthmatics. It would be even better, however, to find the cause of the illness, so that asthma can be totally avoided.

"....I have been close to death twice with asthma and now I'm finding that I can clear the mucus from my lungs much more easily. I also feel an energy surge within one minute of drinking the beverage."

" ....I must say that Kombucha picks me up. I have asthma from working in a clothing factory for years."

THE BLADDER

Dr Herrmann researched whether Kombuchal (Kombucha extract) could dissolve bladder stones. Phosphate stones were inserted into the bladders of male rabbits. Over several weeks, the animals received Kombuchal two or three times daily, and the phosphate stones slowly reduced in size, eventually being discharged through the bladder. An increase in calcium and phosphate could be recorded in the urine prior to the stones disappearing. The Czech scientist discovered that phosphate stones could be largely dissolved in the laboratory with gluconic acid. He believes that glucon acid, taken daily over a period of years, can be absorbed by the human organism. Free glucon acid, chemical or biological, is produced with a Kombucha brew based upon green tea or black tea. (See also *"Prostate Problems"* p.85)

BRONCHITIS

An improvement with bronchitis is often possible through the drinking of Kombucha. A doctor from the Netherlands (Harnisch) said that, after experiencing the benefits of Kombucha for himself, he successfully prescribed it to children suffering from bronchitis.

CANDIDA ALBICANS

Some people suffering from candida albicans only have to think about yeast to have a negative reaction! In an interview with *Search for Health,* Kombucha expert Günther Frank mentions that the Kombucha yeast components help in the fight against candida instead of aggravating it. Split yeasts *(Schizosaccharomyces)* do not contain the spores from which so many people suffer. On the contrary, they can be helpful against damaging yeasts.

It is claimed that the yeasts found in Kombucha will over-run other yeasts, but need time to do so. Candida sufferers should drink the beverage only in moderation and should be careful not to overstrain yeast sediments when bottling the tea. This also applies to non-candida use. This tropical split yeast has the highest fermentation rate in the blood temperature range and can take the place of other yeasts. *(Golz)*

CHOLESTEROL

The media have released many reports on cholesterol, although scientists are not all in agreement about it. High blood pressure, high cholesterol and the resulting blockage of arteries, are regarded as the main cause of heart attacks. Dr Herrmann, with his team of scientists in Prague, conducted animal trials, in which he increased cholesterol levels up to thirteen times above normal. The trials were conducted with the product VigantolR. Dr Herrmann fed cats the deadly dose and, at the same time, gave them Kombucha drops. The result was that the cats were able to survive this dose. He reported that the combination of VigantolR and Kombucha did not lead to an increase in cholesterol levels. Although animal trials cannot always be related to human beings, this test would tend to indicate what people using Kombucha have themselves confirmed. In the Russian clinic of Omsk, a reduction of cholesterol levels was discovered after using Kombucha for a period of three weeks.

CHRONIC FATIGUE SYNDROME (ME)

This troublesome condition is one that many doctors are still reluctant to acknowledge. After the first publication of this book I received many comments from ME sufferers. They asked why I hadn't written about the positive effects of Kombucha on Chronic Fatigue Syndrome, since Kombucha is the only medicine that eases the suffering. Letters from New Zealand and Britain endorsed this thinking. June from Northern Ireland, for example, suffered from aching muscles and low energy for over two years and her symptoms disappeared within three weeks of drinking Kombucha.

Mrs J.E.B. from New South Wales wrote: *"I have been taking Kombucha tea for two months as I have been ill for four years with ME/CFS. When we spoke I had been free of one of the many symptoms, namely headache and fogged brain for five weeks, and there are signs of overcoming the type of Irritable Bowel Syndrome that appears with this illness. However ME/CF Syndrome is characterised by periods of remission and relapses, with relapses becoming shorter, hopefully, and less severe as recovery progresses"*.

Mrs S.H. wrote *"I have had CF Syndrome for some years and have tried many different products i.e., vitamins etc. to no avail, I have found Kombucha products very helpful."*

My observation with CFS is that people suffering from this disease sleep in bedrooms with high earthray and electro-smog pollution and one cannot say that this is merely a coincidence. Electric measurements within the bedroom would, therefore, be strongly recommended, whilst continuing to take Kombucha.

COLDS

It is said that a cold generally lasts about fourteen days when one goes to the doctor and two weeks if treated with favourite remedies at home! In other words, there is absolutely no difference. No one remedy has yet been found to alleviate the uncomfortable effects of influenza. Constantly, new and more resistant viruses make the fight against influenza harder and harder. Even Kombucha users are not resistant to 'flu, although they claim they rarely become ill. Chronic colds, however, have a different cause and should be examined in a little more detail. Often the cause comes from an unhealthy diet and life-style, creating poor bodily functioning — i.e. from a non-healthy digestion. Kombucha has the benefit of being a natural antibiotic and is a recommended treatment during the period of a cold and then continuing to strengthen the system.

CONSTIPATION

Remarkable in almost all reports of treating illnesses with Kombucha is the effect that the beverage has on secondary illnesses, such as chronic constipation. This raises the question, once again, which comes first, the chicken or the egg? It was Hippocrates who said that death sits in the bowels and that a bad digestion is the root of all evil. Modern scientists say that the essential requirement for any healing process is a healthy intestinal flora.

Kombucha, with its living bacteria and yeast components, improves and aids digestion. Other remedies for constipation can lead to diarrhoea if used excessively; Kombucha applies a balance. One's own experiences are frequently the most convincing. I have always suffered from a nervous stomach and intestinal problems, which is one reason why I have become so enthusiastic about Kombucha. Constipation had long been a problem for me, especially when travelling. In 1982, this led to an operation for intestinal closure. All remedies I had previously tried, both commercial and natural, only had limited success. The dose either had to be increased after a short period of time or I had to change to another remedy.

I conducted trials on myself, including drinking 2 or more litres of Kombucha over a period of fourteen days. The results were positive, and alleviated many of my symptoms.

With Kombucha, I now seem to have found the one remedy that really helps — and tastes good at the same time. With constipation it is important to investigate the cause; this could be diet, but it is also true that 'Not everyone needs to go on a daily basis!' Many people tend to panic if they don't have the desired evacuation daily and grab tablets to change this which, in my opinion, is the worst thing anyone can do, and definitely not something one should do on a regular basis. Headaches and constipation are generally the most common symptoms experienced when something is wrong with the body.

The cause, in most cases, is our diet and insufficient exercise. In some cases it may be due to stress or to the medicine one is taking. People who often delay going to the toilet because they do not have the time will programme their bowels to become lazy. The problem for many people can often be solved by merely changing the diet to wholemeal products, vegetables, fresh salads and also natural fat such as butter, cold pressed oil, as well as nuts, etc. Fresh produce should make up at least one third of our diet. One day per week only raw foods should be consumed or a no-food-at-all day (as is practised in many religions) can be made a rule. Drinking a glass of Kombucha in the morning provides a vital aid to the digestive process. Laxatives — including naturally-based laxatives such as senna — should be taken only as a last resort, when nothing else works.

Plum juice and dried plums should also only be consumed for a short period of time. When using laxative teas one should stick to the milder types and change often to avoid any long term ill-effects. The same goes for tea mixtures.

J.A. from North Queensland wrote: *"I have recently been introduced to the miraculous Kombucha and I am ecstatic with results. To date constipation and stomach pain no longer haunt my days."*

DIABETES

In many parts of the world, the Kombucha culture is being passed on to friends in the reassuring knowledge that there are no negative side effects. The question is often raised, however, about the possible effect of the sugar on diabetics. In fact, the fermentation process converts sugar into other products after some 8-10 days. Therefore, as long as the brew is not drunk before the fermentation is complete, with a recommended brewing time of no less than 8-10 days, there should be no problem for diabetics. Pastor Weidinger knew this from first-hand experience as a diabetic himself.

Should diabetics use caution with Kombucha beverages that are being sold in stores? Commercial production may generate ambitious advertising in which all claims may not be met. Where these beverages contain unfermented sugar they must be avoided by diabetics.

Some herbal teas when mixed with green teas are suitable for diabetics; these are Billberry leaf tea *(Vaccinium myrtillus L),* Knotgrass *(Polygonum aviculare L.),* Burdock *(Arctium lappa L.)* and Chicory *(Cichorium intybus L.).*

DIARRHOEA

Kombucha can be used successfully by people suffering from diarrhoea. Tea mixtures of black and plantain tea are often recommended. Scientific trials have been conducted at Omsk in Russia with infants suffering from bacterial dysentery. Only days after the use of Kombucha drops were introduced, the diarrhoea retreated. The weight of the infants increased and so did their appetites. The dysentery bacteria could not be found after a week.

ECZEMA

Eczema can be extremely debilitating and unsightly, causing irritation and embarrassment to those suffering from it. Kombucha has been found to be soothing and healing when the liquid is applied externally to the affected parts several times a day. In addition, drinking Kombucha will strengthen the whole system, working from the inside out. (see also *"Psoriasis"* and *"the Kombucha Bath"*)

FLUID RETENTION

Several reports indicate that fluid retention will decrease with the help of Kombucha. My father (then 87 years old) had to visit the doctor at least twice a week to have the fluid removed from his knees. Through drinking stinging nettle tea he did not require any treatment for the following eight years. Nettle tea can be combined with Kombucha.

A woman in her 50s in England who suffered regularly from swollen ankles (oedema) found, especially after a day's work, that within two months of drinking Kombucha she no longer had the problem, and it has not recurred.

GOUT

Acute gout is only the tip of the iceberg or, as one naturopath has said, "the result of many food mistakes". Intestinal disorders can be genetic, which means that people who have this disease in the family should, by all means, avoid foods which lead to a high level of uric acid. Pain is caused by the build-up of many small crystals of uric acid. In early days, gout was called the illness of the rich. It can be traced back to a rich diet. In the so-called starvation years after the World Wars, gout was one of the rarest of illnesses. Coffee, however, is often named as one cause of gout.

Dr Helmut Golz has written that alcohol contains the component purin which is transformed in the liver into uric acid. Half a gram of uric acid is normally discharged daily by a healthy human being. Uric acid, however, is increased by alcohol intake, which leads to gout. Black tea, also known as Russian tea, has the highest level of purin of any kind of tea, according to Dr Golz. Since only this type of tea was used in Russia and Germany for trials, I assumed that Kombucha would have a negative effect on gout sufferers. However, Dr Golz has explained that the Kombucha symbiosis requires purin for its own digestion, during which uric acid, which is generally hard to dissolve, is turned into an aqueous solution and is easily discharged from the body via the bladder. The Kombucha fungus thus transforms both sugar and purin components during digestion, with the result that they are not harmful to the human body.

HIGH BLOOD PRESSURE

High blood pressure and associated illnesses are at the top of the sickness lists in Western society today. Kombucha is often quoted as a remedy against high blood pressure. One of my friends who suffers from this tried it and some improvement was noted. I recommended fermenting Kombucha with hawthorn tea, and this had better results than those made from hawthorn on its own. Afterwards, the results of blood pressure tests, taken both at home and by the doctor, were that the blood pressure had become normal.

A similar observation was made by another high blood-pressure sufferer, Mrs M.Y. who wrote: *"I have been drinking Kombucha for the past four weeks made with hawthorn tea, to bring down my blood pressure. I take approximately 200ml per day along with two cups of freshly brewed hawthorn tea. A combination of the two teas has proved extremely effective in keeping my blood pressure within the normal range, and I have been able to dispense with prescribed medication. However, because of a previous health problem, I need to be certain that I am doing the right thing and would appreciate any advice you are able to offer in this regard. In your opinion, is it advisable to take so much hawthorn tea/Kombucha over a prolonged period, or should there be 'rest' periods? If my blood pressure should rise during these rest periods, is there an alternative substance I can use?"*

My response is that, first of all, it is strongly recommended that you should never dispense with prescribed medication without consulting your practitioner. This may present some difficulty, since some physicians automatically advise against traditional treatments such as herbal medicine. They advise stopping any herbal treatment in favour of the recommended chemical one. Other more open-minded practitioners may advise their patients with more objectivity.

The recommended dose for high blood pressure is two cups per day; for low blood pressure one cup per day. In the case of Mrs M.Y., I would say that her daily intake of 7oz (200ml) should be either in the form of hawthorn/Kombucha, or two cups of hawthorn tea, but not both. 'Rests' are usually advisable, since the body adjusts to treatments. However, hawthorn seems to be an exception. There are many people who have been taking hawthorn on a daily basis over decades with great results. Nevertheless, there are some warnings to be heeded. Hawthorn was banned in Australia for its alkaloid content.

Overdoses can have negative side effects. As with other remedies, it is the dose which makes the medicine - or the poison.

My father's experience of blood pressure and heart problems provides a good example of what I mean. Born in 1899, he had to join the army at the young age of 17. The military doctor who examined him advised that he should be discharged after a short time of training owing to his heart condition. During the last twenty years of his life, he took hawthorn tea or tincture regularly to control his heart and his blood pressure. When he was hospitalised at the age of 96 and his death was expected, I asked the doctor how much time he had. The doctor replied that the heart was so strong, it was preventing his death!

Mrs. R.H. wrote: *"I thought you would like to know that people with high blood pressure may have to reduce medication while taking Kombucha."*

Experience has shown that people using Kombucha with certain medications may experience a fall in blood pressure; if this is likely they should always consult their physician.

KIDNEY DISORDERS

In his comparative trials with Interferon, Dr A Wiesner found that the biological food Kombucha was 89% as effective. Dr Harnisch reported the case of a sixty year old public servant whose kidneys had not functioned well since childhood. His uric acid level was also too high. No other remedies (including alternative remedies) helped until Kombucha was taken. His kidneys have worked properly ever since.

MULTIPLE SCLEROSIS

In trials, Dr A Wiesner tested the result of Kombucha in comparison to the medicine Interferon. Kombucha showed an 80% effect in comparison to the modern drug. In a letter published in the magazine *Op Zoek*, Mrs M.W. from Holland, who suffers from multiple sclerosis, wrote of her experience of taking Kombucha. In April, 1989, she started to drink Kombucha and experienced complete detoxification. She is no longer tired and has also has got her regular driver's licence back. She writes in her letter, dated February 1990, that, after having taken Kombucha for half a year, she even plans to go skiing again. She wants other people to become aware of the benefits of the Kombucha fungus so that they may experience the same improvements.

PROSTATE PROBLEMS

One of the principal problems associated with male ageing is prostate disorder. Approximately one third of men have prostate surgery and a further one third experience problems of one sort or another. The tea of the alpine willow-herb *(epilobium roseum and epilobium parviflorum)* has been used for many generations in the Alp regions of Europe to counter problems associated with the prostate. This remedy has been available in Australia since 1983, and sold as 'Prostawort'. When combining this with a Kombucha fermentation it should be mixed with 80% of green tea. The herbs should not be more than twelve months old. With inflammation of the bladder this tea has shown some good effects. A mixture of Horsetail, Calendula and Prostawort in equal parts has also shown good results. An elderly lady suffering from bladder inflammation for decades now drinks this tea mixture at the first sign of the problem recurring, eliminating the problem before it sets in.

PSORIASIS

Kombucha is recommended by doctors who have acknowledged this fermented beverage as a help for psoriasis. A woman who suffered from the disease for well over thirty years, during which no other treatment was found to be of help, rediscovered soft smooth skin again after drinking Kombucha.

RHEUMATISM

When compared to Interferon, Dr Wiesner found Kombucha was approx. 92% as effective in the treatment of rheumatism. One of his patients, however, reported that, after having been treated unsuccessfully with other products, she found, upon trying Kombucha, that her fingers were without pain and her normally painful arm could be moved easily. After further taking Kombucha, she was totally free of rheumatism.

SLEEPING DISORDERS

Sound sleep is just as important as eating and drinking. We can survive several weeks without food, but not without sleep. Each individual must determine how much sleep he or she requires. Some require a mere five hours of sleep per day, others need ten. Stress and unsolved problems can be the cause of sleeping disorders along with environmental factors. People having problems sleeping through the night have found that a glass of Kombucha is extremely helpful, while others who derive an initial boost of energy from Kombucha and are prone to bouts of late night activity need to take it earlier in the evening.

STOMACH AND BOWEL DISORDERS

The use of Kombucha, in conjunction with other treatments, has time and time again had the effect of curing stomach and intestinal problems. One might conclude that, because the intestinal flora was reinstated through drinking Kombucha, the cause of the original illness could have been disturbed digestion. As with all uses of Kombucha, the first positive sign is usually a balanced digestive system. The rebalancing of intestinal flora is not only to do with the healing of stomach and intestinal disorders, but is the key to general good health.

TONSILLITIS

In the Russian clinic of Omsk, people suffering from inflammation of the tonsils were treated with Kombucha tea. They gargled with it ten times a day, keeping the fluid in the mouth for up to fifteen minutes each time. Inflammation of the tonsils was reduced and, as with other illnesses where Kombucha was administered, positive side effects could be observed in the healing of intestinal diseases and nasal cavity inflammations.

WOUNDS AND ULCERS

The fungus itself can also be used to heal wounds and ulcers. A woman from Queensland told me that she had suffered for many years from a leg ulcer. She bandaged the fresh fungus direct from the brewing container onto the ulcer which healed in less than a week and has not returned. Other people have placed a young fungus directly on the wound following injuries.

BEAUTY THROUGH KOMBUCHA

The first time I heard miraculous claims being made for Kombucha as a beauty treatment I was very sceptical. Kombucha as a cure for wrinkles was something I regarded as being too far-fetched. Once wrinkles are present, or so I assumed, it is impossible to iron them out with Kombucha. I did think, however, that it might be possible to use it to prevent them forming. As a side effect to Kombucha's help with the digestive system, it should, in theory, be possible to improve unhealthy, rough skin, which often comes from an unhealthy diet or a bad digestion.

I don't believe that drinking Kombucha will banish wrinkles. But don't give up; you can also apply Kombucha externally in the form of an ointment. It is not hard to make. The fungus itself can be pulverised in a few seconds into a cream using a conventional kitchen mixer. Apply the cream to your face, leaving it for twenty minutes while you lie down and rest. Then rinse off with warm water. The cream, with its large content of living yeast components, is also an effective moisturiser. If you consider that the active ingredients of most expensive skin-products are preserved yeasts, you can assume that active and living yeasts should have even better success. With cellulitis, a second treatment, involving massaging the affected areas of skin daily with Kombucha cream, can be tried.

KOMBUCHA AS A POULTICE OR COMPRESS

The author received the following letter: *"A friend gave me a starter 'mushroom' some months back and we've been using it ourselves and sharing it with others. I'd like to tell you about a sixty-six year old lady who has used the Kombucha in rather unusual ways with incredible results.*

"Margaret has been under a great deal of stress for over two years since her husband left her, and this has affected her health severely in a number of ways. She has had constant pain in her left hip and lower leg for many months, and when we arrived to stay with her for a short time recently, we found her unable to lie down, and she was regularly sitting up in a chair at night to try to get some sleep. For her to lie down in bed was unbearable. She also suffers with hypoglycemia and was afraid to take the Kombucha internally, so she decided to try the mushroom as a poultice on her hip. She covered the membrane with 'glad-wrap' [seran wrap/cling film] and some old panty-hose, and within only minutes, the pain began to subside. She gingerly took herself off to bed about 9pm, and actually slept right through to 4am. She was so delighted with the pain relief that she now has Kombucha mushrooms growing in many containers and is continuing to apply the poultice whenever the pain recurs.

"She also suffered from constipation, and so one evening before retiring applied a mushroom to her abdomen. This also proved extremely effective and her bowel movements have returned to normal.

"Margaret also has a very fair skin and had a nasty growth on the side of her face in front of her left ear. A few Kombucha applications and this growth has recently reduced in size and is almost gone.

"This dear lady is no foreigner to natural healing methods, as she is president of a natural therapy group, and has tried many other alternative medical treatments. None has proved as helpful as Kombucha. This very clearly has shown me that this special ferment has incredible potential to help alleviate pain and disease in suffering humanity. I trust you'll find this interesting and helpful."

And this letter: *"...recently I have been having Kombucha tea and noticed a very big improvement (I have had cancer of the mouth for 14 years). Where I had very little voice left, a few weeks on Kombucha tea and my voice is as strong as it used to be. General health seems better too."*

More: *"My husband suffers from rheumatoid arthritis in his right knee. He has had several operations in the past five years. Despite this, his knee is still hurting him, swelling up from time to time and causing great pain. Some days he can't walk or sleep because of this pain. My friend gave me some of the Kombucha fungus she received recently to try on my husband's knee. I applied the fungus directly to the affected knee as a compress. For the first few minutes he felt a sharp pain right across the joint, but the pain gradually dissipated. The fungus had a soothing, cooling effect right away. He was soon at ease and without pain. We left the compress on the knee overnight to see what effect it had. He slept in peace without pain for the first time in months. We will continue the therapy to see if it really will help this debilitating sickness."* [See also "Wounds and Ulcers"]

For the treatment of shingles, eczema and skin fungi, Kombucha compresses are recommended. Packs soaked in Kombucha and applied to the scalp can also help to develop stronger and more beautiful hair.

(**N.B.** Protect bed linen and clothing, as Kombucha tea and the liquid from the fungus does stain.)

THE KOMBUCHA BATH

Pastor Kneipp, one of the most famous natural healers in the German speaking countries, has developed a wide variety of water therapies, which today are being used in six thousand hospitals and private clinics. The human body can easily absorb healing substances through the skin. When added to a bath, Kombucha has good results in energising the whole body and improving general wellbeing. As a general beneficial treatment for common skin problems, add approximately half a litre of Kombucha tea to an average bath.

SLIMMING WITH KOMBUCHA

When talking about slimming, we first have to establish what exactly we mean by slim. The desire to be slim has produced such a 'fat market' for industry, that it is very difficult for the average, slightly overweight person to make the right decision. In 1993, 65.6 million Deutschmarks were invested in advertising by the 'slimming' industry *(Wirschaftswoche, Hamburg Germany, issue 12/1993).* Similar huge figures were spent in the USA, the UK and most other image-conscious western countries where diet, fitness and the marketing of special, calorie-reduced products are big, big business. There is a personal ideal weight for everyone who is reasonably healthy and realistic about their physical image. But many people, especially women, make their lives miserable by setting themselves quite unrealistic goal weights and constantly changing their diets.

New 'miracle' diets are constantly being offered in newspapers and women's magazines. Those who slavishly follow these badly-balanced eating regimes run the real risk of making themselves ill through malnutrition. There are no miracle treatments that can, for example, give an average-sized woman the willowy frame and body structure of a model. Before embarking on any weight-reducing programme, you must ask yourself whether you prefer to be slim and sick or healthy at your personal ideal weight. Being too slim is just as dangerous as being grossly overweight. As with many things in life, the aim should be to find a happy medium.

So what has all this got to do with Kombucha? In the long-term, only a sensible diet can lead to the maintenance of good health and an ideal weight, but Kombucha can have a beneficial effect. For use as part of a sensible weight loss programme, the tea should be fermented for up to fourteen days, by which time it will have acquired a sour taste. A glass of the beverage should then be drunk early in the afternoons and evenings, prior to meals. Once the required weight has been achieved, further brews of the beverage should be fermented for only six to ten days and drunk as part of a balanced diet. Kombucha supports the digestion processes and, therefore, helps to achieve an ideal individual weight. It has a mild purgative effect which is more noticeable when taken before meals. In contrast to laxatives, however, Kombucha does not increase its function with larger dosages.

In an age in which many people do not have enough exercise, have sedentary occupations, eat irregular meals and include much processed food in their diets, Kombucha is especially recommended. A sluggish bowel will lead to constipation, excess weight and many other symptoms associated with an unhealthy lifestyle. The positive effect of Kombucha on health can definitely be related to the effect it has on the regeneration of intestinal flora. Scientists world-wide agree on that point.

When I released my video in 1984, "Back to nature with medicinal herbs", in which I included references to slimming and ideal weights, one famous women's magazine took notice. Following an interview with me, the magazine used the method I recommended in a trial which was successful with every person taking part. However, as I refused to change my report and turn it into a diet plan, the story was never published. Since that time, this magazine and many others have published countless diets — each time contributing to increase in sales from dissatisfied women. However, while any slimming programme however badly-balanced nutritionally and deficient in fresh food, may, in the short-term, lead to weight loss, the long-term results are bound to be ill-health. Health depends on balance. An inadequate diet will never be successful in healing illness or achieving a sensible, permanent weight loss.

KOMBUCHA FOR THE YOUTHFUL OVER 80s

Any report claiming that a 130 year old man and an 80 year old woman were able to have a healthy baby was bound to put the media in a frenzy and draw attention to Kombucha. We can each form our own opinion as to the accuracy of the ages given. However, what should be noted is that many older people believe that Kombucha is the reason for their still-active life. Even Dr Sklenar, who was a military doctor in Russia, reports coming across cases of people of advanced years who appeared to be leading healthy, energetic lives. Kombucha has the effect of detoxifying, aiding the digestion and improving energy. It is not surprising, therefore, that it contributes to the health and fitness of people of advanced years. Kombucha can only be regarded as part of this successful recipe, however, and should not be regarded as its sole cause.

KOMBUCHA AS A REMEDY FOR IMPOTENCE

The potency of a human being is widely-regarded as an accurate measure of health and vitality. In 1992, Germans spent 790 million Deutschmark on sex hormones *(Stern 12/1992)*. That is almost 10 DM per head of population - from new-born infants to great-grandparents. Stress, unhealthy diet, environmental factors, lack of exercise and side effects from medicine can all contribute to a decrease in sexual ability. In his research in Russia, Dr Sklenar found many examples of older people who were still very active sexually. The old farmers in the rural regions he visited attributed this ability to drinking Kombucha. According to the German naturopath Jost Kuessner: "The fungus helps impotence. Many men when they are still young have problems getting an erection. After only one week, Kombucha will help sexual vitality."

HIGH PERFORMANCE SPORTING ACHIEVEMENTS

Members of one of the largest sporting organisations in the world, the Russian military, regularly drink Kombucha. Particular regiments have their own secret recipes for the beverage and the German army has also conducted trials on its effectiveness. Professor Dr Simon Gerrit of the Army Sport School in Warendorf, Germany, tested Kombucha and came to the conclusion that "this pure biological Kombucha fermented tea had a strengthening effect and improved the performances of the athletes".

Russian high-performance athletes are given Kombucha to increase their performance. Scientific tests were conducted in the Olympic centre in Warendorf, with exceptional results. Even with the hardest training there, muscular aches were no problem and the athletes improved their performances through drinking $^1/_2$ US pint (0.25 litre) of Kombucha three times a day. The team of doctors conducting the trials found that salt in the lactic acid *(Lacta readings)* lessened with Kombucha. With high bodily exertion, the organism sours more quickly and Kombucha counteracts this. Overall the athletes were positively affected and recovered much quicker after long runs. Increased wellbeing and performance, according to the scientists, were due to the positive exchange of energy in the cells, encouraged by the Kombucha.

KOMBUCHA USED IN VETERINARY MEDICINE

One man asked me many questions about Kombucha. He mentioned that he was making 30 US pints (15 litres) of Kombucha every day, for his own consumption. I calculated that he must have a family of thirty, based on average Kombucha consumption! But no, it turned out that he had been giving it to his racehorses as an 'energy shot'. "I know that the rewards of Kombucha brewing are 'measurable' in valuable seconds in a race", he said. There are also reports that racing camels in Arabic countries receive a fermented drink as a form of legal 'doping', as the Swiss health magazine *Naturlich*' reports under the heading "Bio-Strath makes camels fast". 73% of Kombucha's production in Switzerland is exported. The ingredients and details of the fermentation are the manufacturers' secret. Many Australian herbalists and naturopaths have spent years ministering to their patients with a secret preparation — which turned out to be nothing other than Kombucha.

Food growers have received more and more criticism in recent years because of their dependence on chemicals. Today the call for natural food is becoming stronger and stronger. Biodynamic produce and alternative healing methods for animals are slowly becoming popular. Kombucha is also being used for veterinary applications. In a trial with sheep and calves, Kombucha drops were applied to animals suffering from diarrhoea with a 100% success rate. With healthy animals, Kombucha was mixed into their food, with a growth increase of 15%.

The economics of Kombucha drinking are, of course, also important. Kombucha fermentation requires a large amount of sugar. Would this not be a beneficial use for the excess sugar we consume and which accounts for a lot of our illnesses? Would it not also be possible to replace health-endangering hormone treatments on animals with a healthy Kombucha product?

A delightful anecdote from an Australian correspondent: *"The most startling success story for me has been with our family pet, a German shepherd dog, who has suffered for the past few summers with an itchy eczema, resulting in a bald patch near his tail; this year some warty lumps appeared. As it is hard to give him the liquid, I began to cut the fungus and gave him a piece each day. He now looks forward to his piece of Kombucha before breakfast. The hair on his back is almost fully repaired, and I expect the warts will also disappear with time; I will continue until they do. He also seems to be a whole lot stronger - his back legs which were becoming weak have strengthened too."*

NETWORKING – GETTING & SHARING YOUR FUNGUS

The way that the Aids community in California has discovered a sense of compassionate sharing of the fungus is a wonderful example of the networking magic of Kombucha. There is probably no other organism that reproduces itself so prolifically every week. It seems wasteful to throw it away when it could bring healing to a friend, or to some stranger who could become your friend. Just as Mother Nature has given us this priceless gift of healing, you can pass it on to someone else.

An excellent way to network Kombucha is to keep a record of those to whom you give cultures. Then, when others ask you for one and you don't have one to spare, simply refer them to someone on your list. Try putting up a notice in your health food shop. Just imagine how our society could be transformed if people got into the habit of growing a culture in their kitchen, just like they used to in deepest Russia!

You don't have to be a mathematician to work out that if everyone you gave one to in a year did the same, there would be thousands of Kombucha cultures out there helping to heal other people by the end of the year! If you all then shared recipes and exchanged tips on ways of making the brew more interesting or of turning it into ointments, what a community builder this would be. Just what we need!

The word about Kombucha is spreading like wildfire. Thousands of people are cultivating cultures and passing them on. If you would like to obtain a fungus, keep your ears open; look in alternative health magazines or ask holistic therapists.

For suppliers of a healthy fungus there are contact addresses in *Appendix Two, "How to Obtain Kombucha" p.105*. Some countries have well organised networks, notably the UK. (Please don't write to the publishers for a fungus!)

EPILOGUE

HEALTH IS BALANCE

Without exception, all Kombucha researchers recommend it as an overall healing method. One cannot isolate certain illnesses and their treatment if holistic health is to be restored. In recent years, there has been a significant resurgence of interest in Kombucha. It should not, however, be regarded as an isolated healing remedy, but rather as what it really is, a very precious, living and health-giving food. If one asks the elderly to what they attribute their good health and fitness, the usual answer is a quizzical look. If one presses further, however, and asks specific questions, the answer that often emerges is, "to a healthy, balanced diet". Kombucha can certainly be a wonderful addition to this.

THE OLDEST 'NEW AGE'

Countless 'New Age' healing methods can be found all over the world but many often become muddled in their migration to other countries. Releasing energy blockages with Reiki, experiencing aura diagnosis using Kirlian photography or plasma print; unlocking fears with 'past-life regression'; mastering the present and future with numerology or astrology; healing with sound or colour, aromatherapy, acupuncture, crystals - these are just a few examples from an endless list of healing aids. Many of them represent the rediscovery of traditions which go back thousands of years. Better communication has spread this knowledge rapidly. No single tradition or practice should be dismissed out of hand, but needs to be considered among the multitude of healing methods available, including modern medicine.

PLACEBO — THE MIRACLE HEALING OF MODERN MEDICINE

The phenomenon of the placebo effect does not fit into the clear picture of natural science. If modern medicine were to use words from the general vocabulary, they would, perhaps, describe it as the 'miracle effect'. Healing with **nothing** is scientifically impossible. Healing with a word or a touch is something we are only familiar with from the Bible. During extensive testing of modern medicine, however, we frequently come across the self-healing or placebo effect. Did science rediscover miracle healing in these trials and did they prove it? Even though placebo medicine does not have any active ingredients, it has remarkably high healing successes for all types of illnesses. Even in 'double blind' testing, the placebo effect has been identified.

For approximately fifty years, research has been testing the effectiveness of placebo medicine. In tests, one group of patients is given a new medicine while, at the same time, a comparative group is given a similar-looking product but without any active ingredients. In double blind testing, neither the patient, nor the doctor know which drug is being taken. If, for example, two thirds of patients who suffer from strong headaches respond to the real drug while the same number of patients in a second group react identically to the 'pretend' drug, this result would not lead to the conclusion that the active component was functioning. Everyone wants to have a local anaesthetic at the dentist. How would a patient react, however, if after treatment, the dentist revealed that the 'anaesthetic' used was only water? About 30% of patients show the same result with a placebo as they do with an actual drug.

The healing effect with a placebo depends to a great extent upon the person administering it. A confident administrator gets a better result than one less self-assured. It is interesting to note that negative side effects from using the real medicine may also be experienced with the placebo medicine. Evidently, the brain is capable of giving healing messages to the body. When we receive medicine, we believe in its success for healing and, therefore, subconsciously heal ourselves. If the self-healing potential of the mind is so powerful, should we not practise this power in order to heal ourselves without any negative side effects? After I published the first edition of this book, I received many letters and telephone calls. The feedback in the context of placebo medicines was particularly strong.

A nurse who, for many years, worked on hospital night shifts told me that it was common practice to give patients a placebo. Patients first received a regular dose of medication to help them sleep but many later asked for a second pill in the middle of the night. The maximum dose permitted per night was one tablet, so the nurses gave those patients requesting a second dose a vitamin C pill instead (visually identical to the sleeping pill) - with a 100% success rate! When I questioned this high success rate, on the grounds that vitamin C has a stimulating effect that is exactly opposite to the result one would expect from a sleeping pill, the nurse only replied, "It simply worked, that's it". My question as to why the safer vitamin C had not, therefore, been given to the patients in the first place was not answered. I can only repeat my question: Why are the millions of dollars which are spent every year on researching new pharmaceuticals not used in the same proportion for researching the possibilities of increasing the 'placebo effect' so that fewer pharmaceuticals would be needed? "When you take a drug with a glass of pure water to swallow it, the water is often doing you more good than the medication." *[F.Batmanghelidj, MD]*

FASTING FOR HEALTH - INSTEAD OF SUICIDE WITH A KNIFE AND FORK

Dietary problems are occurring as never before. We see on our television screens that countless people are starving to death daily in many countries while, on the other side of the world, many people are suffering health problems from eating too much. The question now is, which death is preferable - rapid starvation or slow death through overeating? It may sound strange, but fasting must be regarded as one of the better healing methods. The positive health effect of fasting was recognised by many early religions. One complete day of fasting, or a time of fasting over some weeks with only limited food, is found in most religions. Animals also fast themselves healthy, i.e. they fast during long periods of exertion. The eel and the salmon both fast when travelling long distances to their mating grounds. Migrant birds travel thousands of kilometres without taking any food. Fasting is a highly effective tool towards preventing illness.

Intestinal illness, excessive blood fat and high blood pressure are the main risk factors in the Western world, all of which could easily be treated along with the problems of the Third World countries. If the Western world were to supply the food saved from one day of fasting to Third World countries, there would be no people dying from overeating on one side of the world and no people dying of starvation on the other.

TESTIMONIALS

Thank you everyone who wrote to me about their experiences of drinking Kombucha. The following is a sample:

This is from an Australian medical doctor: *"I have only been using Kombucha for the last 6 months, and have quite a few patients trying it out. So far the main results coming to hand are as follows: 1. Energy increase. 2. Weight loss. 3. Regular bowel motions. 4. One patient with outstanding stomach problems is now 100%. 5. Moles diminishing in size. 6. Good results with sinus and mucus elimination. 7. I used it during the hay fever season and found relief much earlier than usual. We now have 50 families taking Kombucha.*

.... Have recently been introduced to the miraculous Kombucha. I am ecstatic with the results to date regarding constipation and stomach pain which no longer haunt my days.

.... I must say that Kombucha picks me up. I have asthma from working in a clothing factory for years. I have had chronic fatigue for many years and, after a couple of bottles of Kombucha, I felt a marked difference.

.... I have had chronic fatigue for some years and have tried many different products i.e. vitamins etc, to no avail. I have found Kombucha products very helpful.

.... It has become quite 'a craze' at the high school here among the teenage girls who are growing fast and always tired, and they say it has given them more energy — many thanks.

.... I have been buying this medicine (Kombucha) from a naturopath for eighteen months and it has saved both my husband and myself from dying."

"....... Kombucha has helped my arthritis, my writing hand is now pain-free.

"....... I thought you would like to know that people with high blood pressure may have to reduce medication while taking Kombucha, as a lady's blood pressure went too low, taking the medication and Kombucha.

"....... For myself, I am greatly impressed with this tea (Kombucha). I am not young, 77. My digestion is much better and I have far more energy. A friend of mine, a couple of years younger than

myself, has been able to cut her blood-pressure tablets by half. The side effects of the tablets have eased. A highly irritating rash all over her body is now disappearing. She has also effectively rid herself of a clot in her leg by applying Kombucha compresses for three nights. She has always been subject to these clots, and they can last three or four weeks and be the cause of much pain. If I hadn't seen your book in a health shop here in M., these positive results would not have come about. So thank you for writing the book & acquainting us all about this wonderful Kombucha tea".

"....... The benefits are great with Kombucha. Even after a week I'm feeling much better — regular bowel movements, more energy, sleeping well;

.... I have been drinking the tea for about a month and have noticed a change in my varicose veins on my right leg;

.... a 79 year old man has taken tea for two months, has circulation back in his hands; they were black and blue, and are now a much healthier colour; he had a tendency to retain fluid, now he is slim with trim ankles;

Some typical comments about the varieties of taste experienced:
"....... delicious; .... my children thought it was apple juice; .... I like Kombucha more then beer or wine and I feel much better the next day; .... I have brewed Kombucha for a year and am still finding new delicious recipes; really healthy and, at the same time, delicious."

Some more comments about the effects after drinking Kombucha:
" .... cancer growth in the lung shrunk; .... noticed grey hairs on chest and tummy turning black again; .... my blood pressure went back to normal; .... brown spots on skin disappearing. .... urine became clear, from being very cloudy; .... better sleep within two weeks; .... pain relief from arthritis in shoulders and neck; .... my hair thickened up considerably, with less loss when shampooing; .... constipation was a problem for twenty years; with Kombucha everything is just fine; .... urine cleared, weight easier to maintain; .... I simply feel so much better; .... increased energy and mobility playing tennis; .... bowel regularity; .... spots from the sun .... disappearing; .... hot flushes disappeared; .... her premenstrual anger and irritability went; .... not tired any more; .... Kombucha gives me energy; .... the only thing which balanced my digestion; ....

soft smooth skin again; .... I only feel well with Kombucha; ....
Kombucha is my goodnight sleep drink; ....my kidney functions
have reverted to normal; .... diarrhoea retreated; .... reduction of
cholesterol; .... no irritated bladder any more .... "

APPENDICES

Appendix One: **BIBLIOGRAPHY**

Aihara, Herman, "Acid & Alkaline", George Osawa Macrobiotic Fndtn, 1544 Oak St, Oroville, CA, (1980).

Biser, Sam. Interview with F.Batmanghelidj, MD "The Greatest Health Discovery in the World".

Breuss, R. "Cancer and Leukemia - Advice for the prevention and natural treatment of numerous diseases": Bludenz Vgb, Austria, (1982).

Brown,A.J. "On an acetic ferment which forms cellulose": *Jour.Chem.Soc.Lond.* (1886)

Carstens, Dr V. "Hilfe aus der Natur - meine Mittel gegen Krebs": *Quick* (43/1987)

Cribb, A.B. & J.W. *Wild medicine in Australia:* p.60 (1986)

Das Beste, "Geheimnisse und Heilkraefte der Pflanzen".

Fasching, Rosina, *Tea Fungus Kombucha,* Ennsthaler, Steyr, Austria

Filho, L.X., Paulo, E.C., Pareira E.C. & Vincente, C. "Phenolics from tea fungus analysed by performance liquid chromatography": Phyton, Buenos Aires (1985)

Frank, Gunther, W. *Kombucha - Healthy beverage and natural remedy from the Far East:* Ennsthaler, Steyr, Austria

Gadd, C.H. "Tea Cider": *Tea Quarterly* (Talawalelle, Ceylon, 1933)

Gold Coast Bulletin: "Cancer plea bears fruit" and "Pawpaw's medicinal qualities": Weekend Bulletin (1993), P.O.Box 1, Southport, Qld 4215

Golz, Dr. *Kombucha Ein altes Teeheilmittel schenkt neue Gesundheit.*

Goetz, Georg "Kombucha - der Wunderpilz, der Millionen Gesundheit schenkt" in *Das Neue* (Issue 3 - 14, 1988)

Harnisch, Dr Guenther: *Kombucha geballte Heilkraft der Natur"*

Harris, R.D. *Prostawort or Willowherb:* Candelo, Crowsnest, Australia

Hess, Walter, "Bio-Strath macht auch Kamele schneller" in *Naturlich*, AT Zeitschriftenverlag, Bahnhofstrasse 39-43, Aarau, Switzerland (1994)

Horstkorte, C. "Zaubertrank aus China-Pilz hilft auch bei Sex Problemen" Kaminski, A. "Aerzte: Pilz heilt Frauenleiden": *Bild der Frau* (2/1988) Springer Verlag, Hamburg, Germany.

Kanuka-Fuchs, Reinhard. Building Biology & Ecology Institute of New Zealand, 22 Customs St.W., Auckland, New Zealand.

Koerner, H., "Der Teepilz Kombucha": *Der Naturarzt* 108 (1987), and "Kombucha - Zubereitung wurde von Sportmedizinern getestet" *Natura-med* (10/1989)

Lassak, E.V. & McCarthy T., *Australian Medicinal Plants*

Meixner, Dr A., *Pilze selber zuechten:* Aarau, Switzerland

Mulder D., "A Revival of Tea Cider": Tea *Quarterly,* Talawakelle, Ceylon, (1961)

Perko J., "Kombucha - Health you can drink" and other information. Kombucha House, Canungra, Qld, Australia

Potter's *New Cyclopaedia of Botanical Drugs and Preparations*

Quinn, D., *Left for Dead:* Quinn Publishing Co., Minneapolis, U.S.A.

Reiss, J., "Der Teepilz und seine Stoffwechselprodukte": *Deutsche Lebendsmittelrundschau* (9/1987)

Search for Health: "Kombucha 'yeasts' fights candida, they do not encourage it" and "Kombucha Converts Tea and Sugar into a Healthy, Nutritious Detoxifying Beverage".

Sharma, Dr P.R., *The Art of Spiritual Living:* The Ram-Rukimini Institute Liaison, Geneva, Switzerland

Tietze, Harald, "Back To Nature With Medicinal Herbs" (Video 1984)

Tietze, Harald, *Earthrays the silent Killer:* P.O.Box 34, Bermagui South, NSW 2546, Australia

Urban & Schwarzenberg, Roche *Lexicon Medizin*

Vogel, Dr A., *Der kleine Doktor:* Teufen, Switzerland

Waal, de Dr. M., *Medicinal Herbs in the Bible:* Weiser, York Beach, Maine, USA

Wagner, H., *Gegen jede Krankheit ist ein Kraut gewachsen:* Ruhland Altoetting Weidinger, Herman Josef "Kombucha - Tee der aus dem Meere kam" and "Die Kombucharunde" in *Ringelblume:* (1988) 3822 Karlstein, Austria

Wiesner Laboratories: "Kombucha nach Dr med Sklenar" (1987) Schwanewede

Willner, Robert E., *The Cancer Solution:* Peltic Pub.Co., 4400 North Federal Hwy., #210, Boca Raton, FL 33431, USA (1994)

Zimmerman, W., "Wogegen hilf der Kombucha Pilz?" *Fortschritte der Medizin* (12/1989)

Appendix Two: **HOW TO OBTAIN KOMBUCHA**
Suppliers and Further Information

Britain: If you want to acquire a fungus, write to: The Kombucha Network, PO Box 1887, Bath, BA2 8YA **(Please send a stamped, addressed envelope).** The Network was set up by a team of committed people each of whom co-ordinates a specific area. They network Kombucha for their costs only.

For **commercial beverage**: Kombucha Pool of Life UK Ltd supply ready to drink Kombucha in one litre bottles. Made with Japanese Bancha green tea for its health-giving properties. They also supply this tea for home brewing in packets of 125/250/500 grams. For details of outlets, mail order or trade enquiries: Tel: 0171.837.0308.

Kombucha Electric Heating Trays: Give constant warmth, ideal for Kombucha brewing and making healthy baby funguses. Available in two sizes, for smaller batch brewing and for the continuous fermentation method. Safe and economical to use. For further information contact Kombucha Supplies U.K., The Hollies, Wellow, Bath BA2 8QJ *(see advert p.112).*

Kombucha pH Test Strips: Establish the correct degree of fermentation to ensure the greatest health benefits from Kombucha. To order, contact Kombucha Supplies, U.K., The Hollies, Wellow, Bath BA2 8QJ *(see advert p.112).*

The Authentic Continuous Fermentation Jar: Made in the U.K. with an attractive ceramic design and 24 pint/12 litre capacity (based on the jar shown on p.27). The jar is fired with a top quality resistant non-lead glaze. It has a cotton material cover and stylish wooden tap and comes with a complete guide to the continuous fermentation process. For information send an s.a.e. to: The Feeling Good Factor, 36 Braemar Avenue, Wood Green, London N22 4BY. Tel: 0181 889 1426.

Paw Paw extract, for cancer treatment, is available from: The Feeling Good Factor (address above).

Herbs: Baldwin stock the largest range in the U.K. of herbs, roots, barks, tinctures and fluid extracts. Tel: 0171.703.5550 for price list, or call at 171–173 Walworth Rd, London SE17 1RW.

Water filters: Roger Sanders has researched filters extensively and will recommend the most effective and sensibly priced one for your needs. For further information contact: The Well-Being Research

Centre, Westend Farm, 5 Far St, Wymeswold, Leics. LE12 6TZ. Tel: 01509.880.447.

Water Regeneration: Add life energy to water and other liquids by putting it through a mini-vortex. The spiraliser fits in a plastic funnel and costs £7. For further information send a s.a.e. to Aquarian Angel Services, 23 West Mount, The Mount, Guildford GU2 5HL. Tel & Fax: 01483.572.688.

USA: Suppliers of bottled Kombucha brew, and of Kombucha cultures: Lee Vinocur, PO Box 81, Palm Springs, CA 92258 (tel/fax: 619.329.9813); and the Kombucha Hotline, P.O. Box 19037, Encino, CA 91416. Tel: 1 (800) 566.3610.

Germany: Günther Frank, Genossenschafts Str.10, 75217 Birkenfeld im Schwartzwald, Germany.

Austria: Rosina Fasching, PO Box 98, A-9021, Klagenfurt, Austria.

Switzerland: Schweizerische Kombucha-Vertriebs-Zentrale, Postfache 135, CH-8600 Dubendorf, Zurich. Tel: 01.55.69.71.

Kombucha Drops & Tincture: Dr.med.Sklenar Bio-Produkte GmbH, Mausegatt 8-12, D-4630 Bochum 6/Wattenscheid. Tel: 02.327.10075

In **Australia & New Zealand:** The author supplies Kombucha cultures, herbs & green teas, temperature control units. He has fact sheets and runs courses on traditional healing, Kombucha brewing, growing and using herbs, dowsing, building biology, Kneipp applications. Please enclose a self-addressed stamped envelope to: 'Kombucha', c/o Harald Tietze, P.O.Box 34, Bermagui South, NSW 2546, Australia. (Tel: +61-64-934-552, Fax: +61-64-934-900).

　　Kombucha and Paw Paw Leaf extracts supplied by: Kombucha House, Lot 7, Climax Court, Canaugra, Qld 4275, Australia (Tel: +61.75.435.104; Fax: +61.75.435.263).

　　Homeopathic Kombucha: Tina White, Manning Natural Healing Centre, 216 Victoria St, Taree, NSW 2430, Australia.

Appendix Three: - **OTHER BOOKS AND SOURCES**

TIETZE PUBLICATIONS, available in Australian health shops or direct from the author: Harald Tietze, PO Box 34, Bermagui South, NSW 2546, Australia. (Tel: +61.64.934.552, fax: +61.64.934.900)

Water Medicine: — by Harald Tietze.
A practical manual on different water treatments for many health conditions. Includes live and activated water, urine therapy, steam and ice treatment, Kneippp water cure. *100pp Aus$9.80.*

Earthrays: *The Silent Killer* — by Harald Tietze.
The study of the geopathic stress from the Earth's natural energies gone berserk is rapidly providing new insights for traditional health care in the increasing threat of modern diseases such as cancer, asthma, arthritis, rheumatism and more. *100pp Aus$9.80.*

Back to Nature with Medicinal Herbs. *Video by Harald Tietze*
Practical hints for growing medicinal herbs, home made health teas, how to make your own tinctures and herbal ointment. Examples for successful self-treatment with medicinal herbs and biological plant protection against insects. *Available in VHS and Beta.*

GATEWAY BOOKS explore in depth, alternative world views and life scenarios. Here is a selection to whet your appetite:

You Don't Have to Feel Unwell: *Nutrition, Lifestyle, Herbs and Homeopathy* — by Robin Needes.
Do you ever feel you're going through life firing on three cylinders? It doesn't have to be that way. This plainly-put intelligent home guide examines a host of disorders — some common, some serious — and recommends treatment based on a combination of diet, supplements, herbs, homeopathic remedies and sensible living.
336pp, £10.95, $18.95 US

Hormones in Harmony: *A Natural Guide to Womens' Fertility — the Pill, PMT and HRT* — by Robin Needes.
Women have become guinea pigs for new chemical biotechnologies from the contraceptive pill to treatments for PMT and HRT. This self-empowering book encourages women to look at their lifestyles, to use the many tools that nature provides, including homeopathy, herbs, nutrition, flower essences and essential oils, to take control of their own bodies and lives.
224pp, £8.95, $14.95 US

The Golden Fountain: *The Complete Guide to Urine Therapy* — by Coen van der Kroon.
One of Nature's oldest remedies, urine is naturally sterile, and has great healing properties, either drunk while fresh or applied to the skin for wounds or burns. Your own is best! Has had success with HIV patients.
160pp, photos & illus. £8.95, $14.95 US

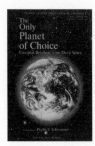

The Only Planet of Choice: *Essential Briefings from Deep Space* — by Phyllis Schlemmer & The Council of Nine.
".....a provocative mind-opening experience" Dr James Hurtak, Academy for Future Science, USA. Gateway's bestseller, compiled from nearly 20 years of communications with a circle of high level universal beings, is widely acknowledged as one of the most significant books for our time. For anyone who wants to know more about the place of Planet Earth in the Cosmos, the origin of humanity, ETs, God, the present traumas facing the world and how they can be resolved. *"Everything you ever wanted to know about the Universe, but didn't know who to ask"* Kindred Spirit magazine, UK. 352pp, £9.95, $15.95 US

The Cosmic Connection: *Worldwide crop formations and ET contacts* — by Michael Hesemann.
The first book to bring together the two most mysterious and unexplained phenomena of our time - crop circles and ETs, showing fascinating connections between them. A global and up-to-date survey of the most convincing information, with many color and black and white photos.
232pp, £12.95, $19.95 US

When the Earth Nearly Died:
Compelling Evidence of World Catastrophe, 9,500BC — by D.S.Allan & J.B.Delair.
Evidence from many disciplines, traditions and cultures, of a cataclysm which nearly destroyed Earth and Mars about 11,500 years ago. The authors draw on decades of research to describe how a golden age disappeared with appalling devastation, and show how their findings could have relevance for present world changes. 384pp, incl. many photos, tables, maps & charts, *£12.95, $19.95 US*

A Search for the Historical Jesus:
From Apocryphal, Buddhist, Islamic and Sanscrit Sources — by Prof. Fida Hassnain.

Millions of people have been brought up to believe that Jesus's life mission ended with the crucifixion. Here, a respected Sufi historian finds evidence of information suppressed by the Church that Jesus survived the Cross and undertook an Essene-backed extended ministry in India and the East. *Riveting reading. 268pp, incl. many photos & maps, £8.95, $14.95 US*

Living Energies: *Viktor Schauberger's Brilliant Work with Natural Energy Explained* — by Callum Coats.
Schauberger was a pioneer in working closely with nature. This important book follows the development of his revolutionary ideas on water purification, transport, free energy heating and home power generation. It lays the foundations for new technologies which can save this planet from destruction. *320pp, 420 diagrams & illus. £13.95, $19.95 US*

Safe as Houses?: *Ill health and electro-stress in the Home* — by David Cowan & Rodney Girdlestone.
To help you identify sources of electrical stress in the home - from nearby electrical generators to leaky microwave ovens and circuits; also covers problems of geopathic stress related to earth energies, and gives instruction and background to dowsing the energies. *224pp, many illus. £8.95, $14.95 US*

In the **USA** Gateway's books are available through many bookstores. In case of difficulty you may contact our distributors: National Book Network, 4720 Boston Way, Lanham, MD 20706 (tel: (301) 459.3366; fax: (301) 459.2118), or Gateway Marketing: 18900 Olive Ave, Sonoma, CA 95476 (tel: (707) 939.1953; fax: (707) 938.3515)

In **Canada,** Gateway's distributor is Temeron Books, #210, 1220 Kensington Rd NW, Calgary, Alberta T2N 3P5. (tel: 403.283.0900; fax: 403.283.6947)

Australia: Banyan Tree Book Dist., PO Box 269, Stirling, S.A.5152. (tel: 8.388.5354; fax: 8.388.5365)

New Zealand: Peaceful Living Publications, PO Box 300, Tauranga, N.Z. (tel: 7.571.8105; fax: 7.571.8513)

South Africa: Wizard's Warehouse, PO Box 3340, Cape Town 8000. (tel: 21.461.9719; fax: 21.451.417)

Singapore: Pansing Distribution, 8 New Industrial Rd, Singapore 536200. (tel: 382.0488; fax: 281.5277)

Index

Kombucha Supplies

KOMBUCHA ELECTRIC HEATING TRAY

Provides low wattage controlled heat with the ideal constant liquid temperature of approximately 24°C/74°F.

- Simple and economical to use
- Cuts down on Kombucha fermentation time
- Foolproof fermentation
- Produces healthy new funguses
- Safe for glass and ceramic containers
- Easy to clean
- 2 year guarantee

Kombucha electric heating trays are available in two sizes:
TE2 – suitable for 1 or 2 batch brewing bowls
TE4 – suitable for 1 – 3 batch brewing bowls; also ideal for the continuous fermentation process where a larger liquid volume is used.

Available with standard U.K. voltage and plug. Can be supplied for U.S.A. and Euro voltages.

KOMBUCHA pH TEST STRIPS

The full benefit of the Kombucha brew is only possible when the sugar has been transformed into health-giving organic acids, enzymes and vitamins. The way to establish the correct degree of fermentation is to test the acidity or pH level of the brew.

We can supply inexpensive pH test strips with a range from 1.7 to 3.4. The target for Kombucha is between 2.4 and 2.8.

Available in glass tubes of 100, giving 200 individual tests.

For further details and orders, contact:
Kombucha Supplies UK,
The Hollies, Mill Hill,
Wellow, Bath, BA2 8QJ.
Tel: 01225 833 150   From overseas: (44) 1225 833 150
Fax: 01225 840 012   From overseas: (44) 1225 840 012.
(Overseas orders welcome, English instructions only. In the U.S.A. the Kombucha Test Strips can be obtained from **The Kombucha Hotline** – 1(800) 566 3610)